Gut Microbiota - A Key Player in Overall Human Pathologies

Edited by Natalia Beloborodova

Published in London, United Kingdom

Gut Microbiota - A Key Player in Overall Human Pathologies
http://dx.doi.org/10.5772/intechopen.1005949
Edited by Natalia Beloborodova

Contributors
Adekunle Muiz Mofolorunso, Adnan Iqbal, Alexander Zaborin, Anosh Tahir, Bukunola Oluyemisi Adegbesan, Esther Nkechi Ezima, Geshu Du, Ghizal Fatima, Gideon Olamilekan Oluwatunase, Haihua Xie, Hu Li, Ifabunmi Oduyemi Osonuga, Jia Jiang, John Alverdy, Kiran Nooruddin, Liyan Yu, Maryam Bakhsh, Mi Liu, Natalia Beloborodova, Nikita Fadeev, Oleg Medvedev, Olga Zaborina, Oxana Povarova, Patrick Godwin Okwute, Penghui Lu, Robert Keskey, Rong Luo, Ruquia Noor, Samuel Oluwadare Olalekan, Shulin Xiong, Sihui Cao, Stanislav Sitkin, Stepan Toshchakov, Timur Vakhitov, Tingxia Lv, Ustin Zolotikov, Weiai Liu, Xiayun Li, Zonggen Peng

First published in London, United Kingdom, 2025 by IntechOpen
IntechOpen is the global imprint of INTECHOPEN LIMITED, registered in England and Wales, registration number: 11086078, 167-169 Great Portland Street, London, W1W 5PF, United Kingdom

For EU product safety concerns: IN TECH d.o.o., Prolaz Marije Krucifikse Kozulić 3, 51000 Rijeka, Croatia, info@intechopen.com or visit our website at intechopen.com.

British Library Cataloguing-in-Publication Data
A catalogue record for this book is available from the British Library

Gut Microbiota - A Key Player in Overall Human Pathologies
Edited by Natalia Beloborodova
p. cm.
Print ISBN 978-1-83635-262-4
Online ISBN 978-1-83635-261-7
eBook (PDF) ISBN 978-1-83635-263-1

If disposing of this product, please recycle the paper responsibly.

Meet the editor

Natalia V. Beloborodova is affiliated with the Federal Research and Clinical Center of Intensive Care Medicine and Rehabilitology, where she serves as chief researcher and head of the Laboratory of Metabolism in Critical States. She completed her higher medical education at Pirogov's 2nd Moscow Medical Institute, Faculty of Pediatrics, Moscow, Russia. She earned her Ph.D. in 1984 from the State Medical Academy of Postgraduate Education, received academic certification in clinical microbiology and chemotherapy, and obtained the academic degree of Doctor of Medical Sciences in 1996. She was awarded the title of Professor in 1998. Natalia V. Beloborodova has been a member of the European Academy of Natural Sciences since 2003 (Hannover, Germany), serves as an expert for the Russian Science Foundation since 2017, and received the title of Honored Worker of Sciences of the Russian Federation in 2021. She is married and has two sons.

Contents

Preface

The term "gut microbiota" has been used in the scientific literature for over 50 years, and, according to the PubMed database, more than 100,000 articles have been published on this subject. However, more than 95% of articles on this topic have been published over the past 10 years, reflecting the explosive growth of interest in the role of the microbiota in human pathology. The title of the book highlights the crucial role of the microbiota in various pathological processes; however, this aspect of the microbiota is still rarely considered in clinical practice. Over the past decade, significant advancements have been made in our understanding of the problem's complexity. Several chapters in this book discuss new concepts and previously unknown mechanisms of the relationship between the gut microbiota and human pathologies.

Today, the gut microbiota should be considered a full-fledged, multifunctional organ whose cells (prokaryotic microorganisms) differ from the eukaryotic cells of the host human, but work in harmony and synergy in symbiosis with it. In the first chapters of the book, based on experimental and clinical studies, the detailed mechanisms of interaction between the microbiota and the macroorganism are shown to be disrupted, contributing to the development of life-threatening infectious complications, such as those occurring during high-risk surgical operations.

It is essential to consider the risk of microbiota disruption resulting from the excessive use of anti-anaerobic antibiotics, which can lead to the transformation of the microbiota into a pathobiota, causing severe metabolic disorders, organ dysfunction, and sepsis. On the other hand, the authors demonstrate how well-founded measures aimed at preserving the beneficial microbiota offer new perspectives in the treatment of life-threatening conditions.

The gut microbiota has a significant impact on both physical and mental health. The book discusses the relationship between the microbiota and various age-related cognitive disorders, including Alzheimer's and Parkinson's diseases. The authors explore how gut-targeted therapies can be a promising approach for improving cognitive health and optimizing memory function. The book also discusses bidirectional regulation in the microbiota-brain system, where signals from the brain regulate nutrient intake, gut motility, and secretory activity to maintain the microbiota's balance, while hormones, cytokines, and other signals from the gut contribute to the synthesis of neurotransmitters, manage stress responses, mood, behavior, and more.

The book's chapters analyze data on the composition of the microbiota in various patient groups, as well as the results of monitoring using classical microbiological methods, polymerase chain reaction, and sequencing. Gas or liquid chromatography-mass spectrometry, as well as other modern technologies, are used to determine the metabolic products of bacteria and archaea.

The book devotes considerable attention to the treatment and correction of the microbiota, including options for traditional, complementary, and alternative therapies. It presents the results of the effectiveness of the selective decontamination strategy using certain oral antibiotics, as well as objective data from studies on probiotics and metabiotics, including fundamentally new ones, for the correction of microbiota metabolism.

The main conclusion of this book can be summarized by the phrase "A person's full health can only be achieved when they are in harmony with their microbiota". There is still much work to be done to find ways to achieve this harmony, and despite the discoveries and advancements of recent years, we seem to be just at the beginning of this journey.

People from different professions and countries were involved in the creative process of creating the book. I would like to thank everyone who generously shared interesting results and new knowledge on the pages of this book, as well as all the IntechOpen employees who were involved in this project.

Natalia Beloborodova
Professor,
Federal Research and Clinical Center of Intensive Care Medicine
and Rehabilitology,
Moscow, Russian Federation

Section 1

Microbiota Management Strategies in High-Risk Surgery and Intensive Care

Chapter 1

Destruction of the Anaerobic Gut Microbiota in Hospitalized Patients Increases the Risk of Sepsis

Natalia Beloborodova

Abstract

The anaerobic gut microbiota is considered as a kind of multifunctional organ, providing, among other things, protection against infectious complications and sepsis in hospitalized patients. At the same time, the use of some broad-spectrum antibiotics leads to the destruction of gut anaerobes, which correlates with the negative treatment outcomes. Since the 1980s, the selective decontamination (SD) strategy has been known for preserving anaerobes. The preventive use of SD demonstrates a reduction in infectious complications and mortality in high-risk patients, so SD remains relevant today. Sepsis, a life-threatening condition, is often referred to as mysterious because it can develop in any critically ill patient, regardless of the underlying disease. According to the current statistics, sepsis is the leading cause of one in five deaths worldwide and still holds a leading position in the mortality rate of patients in intensive care units (ICUs). A promising way to understand the mechanisms of multiple organ dysfunction and sepsis is through a metabolomic approach. Objective monitoring of clinically relevant microbial metabolites in patient blood has shown that the risk of developing and progressing sepsis is associated with microbiota dysfunction. The authors see the solution to the problem in the preferred choice of antibiotics that do not damage anaerobes and, moreover. create conditions for restoring the metabolism of beneficial anaerobic bacteria in the gut.

Keywords: gut microbiota, sepsis, anti-anaerobic antibiotics, critically ill patients, infection prevention, mortality, intestinal anaerobes, selective decontamination, microbial metabolites, new strategy of antimicrobial therapy

1. Introduction

Currently, according to the SEPSIS-3 concept, adopted by the world community in 2016, sepsis is considered as an organ dysfunction, caused by a dysregulated host response to infection [1]. It is important that recommendations for the diagnosis and treatment of sepsis traditionally assess the functional state of organs such as the heart, lungs, brain, liver, and kidneys, but the role of the microbiota as a complex, multifunctional organ has been left out of consideration. I would like to express my hope and confidence that in the near future, when developing the next conceptual document on sepsis (for example, SEPSIS-4), the microbiota as an organ will be included in the recommendations for monitoring and correction.

During the same period, 2016–2017, numerous studies have been published that convincingly demonstrate the great importance of the gut microbiota in the course and prognosis of critical conditions [2–4]. The researchers call the discovered phenomenon "extreme dysbiosis of the microbiome" in critically ill patients [5].

The research results demonstrate profound violations, a sharp suppression of the species composition and diversity of the microbiota, which in critically ill patients receiving massive antibiotic therapy is often represented by only a few types of hospital antibiotic-resistant strains, when the term "ultra-low-diversity pathogen communities" can be used to describe the gut microbiota [6, 7].

2. Impact of antibiotics on gut microbiota

In the same years, an article by Haak et al. suggested the need for microbiota-targeted therapies in the intensive care unit [8]. The authors noted that the composition of the microbiota undergoes rapid and extreme changes during critical illnesses, losing its ability to perform protective functions for the host body. Moreover, in conditions of elimination of obligate anaerobic bacteria, the gut of patients can be freely populated by hospital bacteria and turn into an additional source of pathogens, such as *Klebsiella pneumonia*, *Pseudomonas aeruginosa*, *Acinetobacter baumannii*, etc., contributing to the recurrent course of infections and sepsis.

The authors searched for answers to the following questions: "how our microbiome is functionally changing over the course of critical illness, and how these pathways could be modified to promote recovery from disease." The effects of probiotics (e.g., *Lactobacillus plantarum*) and fecal microbiota transplant (FMT) were studied, and the capabilities of live bacteria *Clostridium scindens* and *Bacteroides thetaiotaomicron* to perform a protective function against infections caused by antibiotic-resistant bacterial strains were evaluated in experimental models. In mechanically ventilated patients, the effectiveness of chlorhexidine for oropharyngeal decontamination and other approaches was studied [9].

By this time, an increasing number of clinicians realized that antibiotics prescribed to combat a specific pathogen in the treatment of patients with sepsis could simultaneously cause concomitant harm and exacerbate disorders of the species composition and function of the gut microbiota. It is not without reason that the suspicion has arisen that antibacterial therapy itself, by eliminating natural defenses of gut microbiota, can contribute to the excessive growth of resistant microorganisms, the development of subsequent infections and the generalization of the septic process. The search for a differentiated approach to antimicrobial therapy to preserve the obligate anaerobic microbiota formed the basis of the strategy of selective decontamination.

3. Selective decontamination (SDD) strategy

Selective decontamination strategy originated in the 1960s and 1970s based on the concept of colonization resistance of the gastrointestinal tract. Antimicrobial drugs for SD were selected taking into account their high activity against the main endogenous opportunistic microorganisms, but they did not suppress the normal anaerobic gut microflora. We managed to find one of the first publications in the archives (**Figure 1**). In the literature, one can find such a definition of the term SD: "Selective decontamination of the digestive tract (SDD)—the use of (nonabsorbable) antimicrobials that are applied daily in the gastrointestinal tract."

In the 1980s and 1990s, an intensive study of the possibilities of SDD began directly in surgical departments and intensive care units; its clinical effectiveness was evaluated [10]. The preventive effect of SDD was also associated, among other things, with a reduction in the risk of translocation of the gut microflora through mucous membranes, which is extremely important for intensive care patients. A combination of polymyxin (B or E), tobramycin, and amphotericin B was chosen as the first SDD regimen [11].

Experience has shown that the use of only nonabsorbable antibiotics for the prevention of infectious complications is less effective than with mixed SDD regimens, meaning in combination with a parenteral antibiotic [12, 13]. In patients at high risk of developing a gram-negative infection, polymyxin + tobramycin + amphotericin was used enterally simultaneously with intravenous administration of ceftriaxone or cefotaxime. Other regimens have been proposed and investigated, including enteral administration of vancomycin.

SDD has become widely used in the Netherlands for patients suffering from mechanical injury [14]. Then the scope and geography of the method were expanded. The results of many randomized clinical trials (RCTs) on the efficacy of SDD have been published. They have shown the ability of this method to significantly reduce the incidence of infectious complications in patients in the intensive care unit [15]. The results of hundreds of clinical studies and dozens of meta-analyses have demonstrated the advantages of the preventive and curative effect of selective decontamination compared with traditional antibiotic therapy in patients at risk of infectious complications and sepsis, as well as the safety of SDD in terms of the spread of antibiotic resistance [16–18].

Convincing evidence has been obtained for a reduction in mortality in intensive care unit when using SDD in a wide variety of diseases and pathological conditions, for

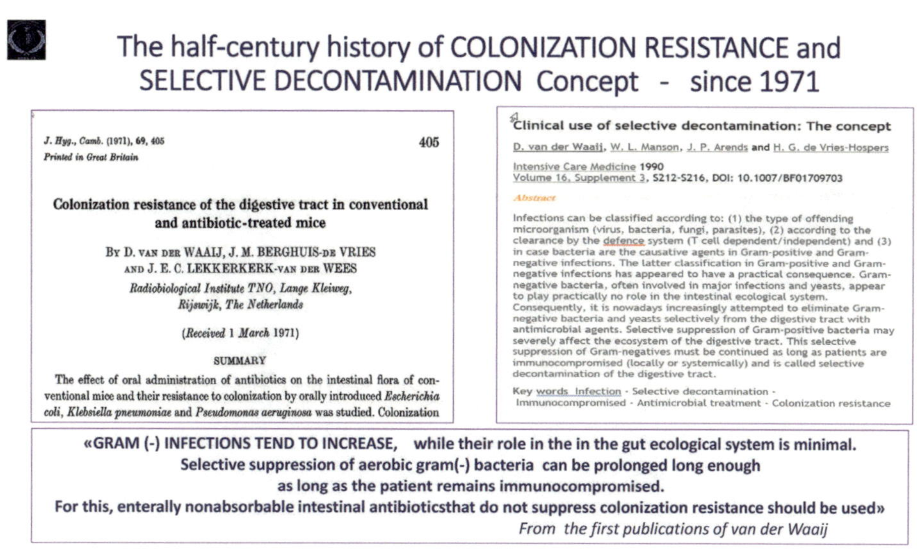

Figure 1.
From the first publications about SDD.

example: during surgical interventions on the organs of the gastrointestinal tract [19, 20], in cardiovascular surgery [21], during liver transplantation, cytostatic therapy, etc. [22].

Today, in 2025, there is every reason to confidently list the proven benefits of SDD, briefly summarizing the accumulated results [18, 23], such as:

- SDD does not interfere with the standard treatment of a patient for a particular disease
- SDD is not inferior to traditional approaches in the quality of medical care
- SDD strategy includes the use of nonabsorbable antibiotics enterally
- SDD reduces colonization of the gut by potentially pathogenic bacteria and fungi
- SDD does not increase the number of antibiotic-resistant microorganisms in patients
- SDD does not contribute to the development and spread of antibiotic resistance
- SDD does not increase the systemic use of antibiotics
- The use of SDD reduces the incidence of nosocomial infections
- The use of SDD reduces hospital mortality.

It is logical that from an economic point of view, the use of SDD is beneficial. For such a conclusion, the last two facts are sufficient—a decrease in nosocomial infections and hospital mortality [18, 23].

Based on the results of a number of large randomized controlled trials involving intensive care unit (ICU) patients. In the Netherlands, most intensive care units

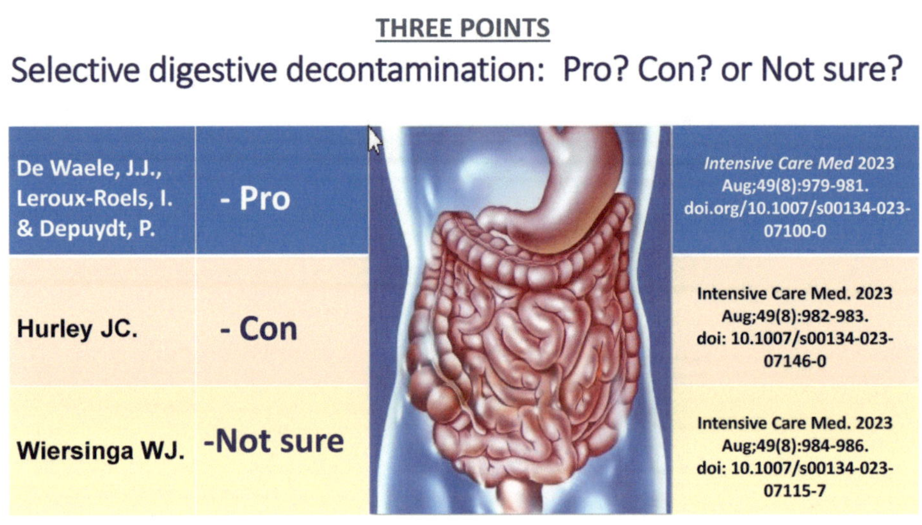

Figure 2.
Three points on CDD: "pro," "con," "not sure."

have switched to SDD, but in other countries, intensive care physicians are reluctant to follow this approach. Some scientists object or still doubt, which is illustrated by the figure with links to three articles that detail all modern points of view on SDD (**Figure 2**).

There are several reasons to explain why there is so much controversy around SDD, which will be discussed below.

4. Anti-anaerobic antibiotics and sepsis risk

4.1 Different antibiotics have different effects on human microbiota

In the coming years, management of the composition and functions of the microbiota in critical patients may be included in the list of life-saving treatment strategies. There are still questions about how to do this to the maximum benefit for the patient. Returning to the issue of the reasons for the disagreement regarding selective decontamination, the most important of them is the choice of the "right" parenteral antibiotic, but this choice is not regulated and is not discussed in the studies listed above. When administering parenteral antibiotics, it is not taken into account that any systemic antibiotic can have a serious effect not only on bacteria circulating in the blood, but also on the intestines. This can have negative consequences, such as the suppression of obligate anaerobic microflora, thereby contributing to the progression of sepsis.

Thus, the effectiveness of SDD in the use of the above-mentioned nonabsorbable antibiotics in a number of patients may be reduced or even completely absent if an antibiotic suppressing anaerobes was used parenterally at the same time. Therefore, treatment results may vary. Unfortunately, it is often impossible to obtain information from the publications about which antibacterial drugs patients received parenterally during and/or after SDD. Hence, the results of treatment may vary.

The effect of intravenous administration of antibiotics on the gut microbiota in cardiac surgery patients was studied. A total of 388 fecal samples from 154 cardiac surgery patients were collected. Various groups of antibiotics were used parenterally, including cefuroxime, ceftazidime, cefoperazone-sulbactam (sulperazone), imipenem cilastatin (tienam), meropenem, levofloxacin, moxifloxacin, ciprofloxacin, vancomycin, teicoplanin, tigecycline, ticarcillin-clavulanate, piperacillin-tazobactam, and penicillin. Thus, using the example of cardiac surgical patients, the authors showed that the intravenous method of administering antibiotics does not protect the microbiota: the results confirmed significant changes in the composition of the gut microbiota [24].

There are studies where the authors analyze the frequency of deaths in the ICU and find a reliable relationship with the frequency of antibiotic use, which can inhibit gut microbiota anaerobes. In a cohort study of 3032 critically ill patients in the Unites States, it was found that systemic anti-anaerobic antibiotics led to decreased gut bacterial diversity and an expansion of *Enterobacteriaceae* spp. For a retrospective analysis, the authors divided all patients into two groups, and analyzed in detail the course and outcome of the disease in the comparison groups. The main criterion for distinguishing the groups was the sign of treatment with antibiotics that suppress anaerobes. In terms of diagnoses and basic clinical and laboratory characteristics, the two groups did not differ in principle. During ICU treatment, patients of the first group (n = 1942), along with other drugs, used such anti-anaerobic antibiotics as piperacillin-tazobactam, metronidazole, meropenem, clindamycin, ampicilline-sulbactam, ceftriaxone; in the second group of patients (n = 1090) these antibiotics were not used. In other words, in this study, in the first group, antibiotics had a negative effect on the patients' anaerobic microbiota, while in the second group the conditions for the preservation and restoration of the microbiota were more favorable. As a result, the length of hospitalization and the survival rate of patients in the second group were significantly higher. The study has convincingly proved that anti-anaerobic antibiotic therapy is independently associated with decreased ventilator-associated pneumonia (VAP)-free survival, infection-free survival and overall survival. The data obtained has been published as a result of a retrospective analysis [25].

It would be possible to assume with a certain degree of certainty that abstinence from taking anti-anaerobic spectrum antibiotics in combination with SDD can actually improve patient outcomes in the intensive care units. We can only hope, this will soon become known from the results of future research. Currently, clinicians do not have the opportunity to assess whether antibiotics prescribed to a patient will have a negative impact on the gut microbiota. The susceptibility testing of anaerobic bacteria recovered from selected cases, which is used for choice of antimicrobial therapy for infections caused by anaerobes, is not suitable for assessing the effects of an antibiotic on intestinal anaerobes. The composition of the gut microbiota is dominated by anaerobes, there are hundreds of species, the final number has not yet been established.

Of the works of previous years, the most famous is a study performed in the 1970s ("at the dawn" of SD) using methods of anaerobic cultivation in test tubes for the purpose of quantitative and qualitative research of the fecal flora of 20 clinically healthy Japanese-Hawaiian men. Were isolated 113 different types of organisms from 1147 isolates, which accounted for 94% of viable cells in feces. This study shows the predominance in gut of the genera *Bacteroides*, *Bifidobacterium*, *Fusobacterium* and *Eubacterium* [26]. Since 2012, a new technique for isolating and cultivating bacteria has been introduced, called "culturomics" [27]. In 2023, it was reported (Wan et al.) that with the help of culturomics technology, the number of bacterial species found in the human intestine reached 3253 [28].

In total, the analysis of more than 2000 samples in six studies of the gut microbiome revealed more than 22 million unique genes in the gut. It has also been found that different groups of microorganisms have functional features, with some of them involved in a wide range of different metabolic processes, including amino acid biosynthesis, while others are involved in degradation processes [29].

4.2 Excessive use of anti-anaerobic antibiotics is not justified

Anaerobes, more often such as *Bacteroides fragilis group*, *Clostridium* spp., *Prevotella* spp., *Peptostreptococcus* spp., *Peptococcus* spp. are involved in the etiology of a number of severe infections. Methods for determining the sensitivity of pathogens of anaerobic infections to antibiotics are well developed and well-known, they are used in specially equipped research and reference laboratories. In clinical settings, if there are indications (abdominal infections, peritonitis, gangrene, abscess, etc.), doctors use recommendations on the choice of antibiotics. The article Brook provides data on 12 antibiotics with a high degree of activity against anaerobic pathogens, among which carbapenems, beta-lactams, combinations of beta-lactams and beta-lactamase inhibitors, metronidazole, and chloramphenicol stand out [30].

In real clinical practice, doctors often exaggerate the need to use anti-anaerobic antibiotics, expand the indications for their use, and use unreasonable combinations of two or more drugs, which can only harm the microbiota, with all the consequences for the patient's body.

There is a growing understanding of the great importance of gut anaerobes for the human body, and more and more scientists are wondering whether we are harming our patients by using antibiotics with an anti-anaerobic spectrum. The researchers are trying to conduct an objective analysis of parenteral antimicrobial therapy, which is prescribed to the vast majority of patients in the intensive care unit. The work evaluates the risk of negative effects of parenteral antibiotics on the gut microbiota, the risk of worsening the prognosis and outcome of critical conditions, in order to determine the optimal treatment strategy in the future.

In one of the articles on this topic, the authors report a single-center observational study that included 192 patients admitted to the intensive care unit with aspiration pneumonia (AsP), community-acquired pneumonia (CAP), and nosocomial pneumonia (HAP). During the study period, the following recommendations for antibacterial therapy were given: both vancomycin and cefepime for HAP, ceftriaxone and azithromycin for CAP, and the addition of metronidazole or clindamycin when anaerobic coverage was desired (AsP). Broader-spectrum antibiotics, which covered both aerobic and anaerobic microbes (i.e., piperacillin/tazobactam or ampicillin/sulbactam), were excluded from the study as there would be difficulty in ascertaining whether the rationale for these antibiotics included treating anaerobic organisms. The results showed that a significant proportion (one in four of these patients) received antibiotics with anti-anaerobic effects, although they did not meet the criteria for receiving them. Inadequate (i.e., excessive) use of anti-anaerobic antibiotics was detected in 12/20 (60%) patients with AsP, 27/107 (25%) with HAP, and 9/65 (14%) with CAP. It turned out that patients admitted to the intensive care unit with pneumonia, who were additionally prescribed anti-anaerobic antibiotics without explicit indications, as a result stayed longer in the ICU [31].

Another larger study (enrolled 2606 patients) also showed that the majority of patients in the hospital, regardless of the presence or absence of aspiration data, received broad-spectrum antibiotics against anaerobic bacteria [32].

Of particular concern are high-risk patients on ventilators in the ICU, in whom antibiotics are prescribed in almost 100% of cases, and not only for treatment, but often for the prevention of hospital-acquired pneumonia. It was shown that in patients treated with anti-anaerobic antibiotics, the survival rate of patients with pneumonia and overall survival were significantly lower. Thus, if there is no attention to the microbiota and in the absence of strategies for preserving anaerobes in the gut, the treatment results of critical patients are worse than they could be in a modern clinic today [25].

It should be noted that in cases of severe pneumonia in hospitalized patients, antibiotic therapy determines the outcome of the patient's treatment. An intensive care physician should be aware of the dangers of anaerobic antibiotics for the microbiota, avoid prescribing them, and use them only when there are clear signs of aspiration pneumonia, as prescribed in official guidelines. Moreover, targeted studies of the respiratory tract did not reveal significant differences in the frequency of detection of anaerobes. When comparing patients with aspiration risk factors and common community-acquired pneumonia (CAP), aspiration community-acquired pneumonia (ACAP) and community-acquired pneumonia in patients with aspiration risk factors

(CAP with AspRFs) are infections associated with anaerobes, but limited evidence suggests their pathogenic role [31, 32].

4.3 Carbapenems are the most dangerous for the gut microbiota

In any case, to preserve the microbiota and prevent sepsis, it should be remembered that anti-anaerobic antibiotics vary significantly in terms of coverage of the spectrum of anaerobic bacteria (**Figure 3**).

Antibiotics kill both "good" and "bad" bacteria; this disrupts the natural balance of the gut microbiome, which plays an important role not only for health but also for the survival of seriously ill patients in the hospital. It is preferable to adhere to a policy limiting the spectrum and duration of use of anti-anaerobic antibiotics. Realizing the potential danger of gross damage to the microbiota, it is advisable to refrain from carbapenems and choose "less aggressive" antibiotics for the gut microbiota. This is especially true for carbapenems.

Research confirms this negative relationship. Back in 1991, more than 30 years ago, when ultra-wide spectrum antibiotics from the carbapenem group first appeared, the authors of this article warned about the negative effects of intravenous antibiotics on the gut microbiota and reported that meropenem is more active than imipenem [33]. The other "signal" about the adverse effects of meropenem on the human microbiota was received by the publishing house in 1992 from Japan in the form of a communication article, but was published only 17 years later [34].

The objections of some authors are based on the fact that in a number of clinical situations, these anti-anaerobic drugs are absolutely necessary, refusal of them can endanger the patient's life, especially since mortality from anaerobic infections is high and reaches 30%. For example, infections such as wound gangrene, perforated peritonitis, deep abscesses (pelvis, brain, abdominal cavity, etc.) are always polymicrobial with the active participation of anaerobes, most often such as *Bacteroides fragilis* group, the share of which, according to the data Park et al., accounted for 41.8% of anaerobic infections, which accounted for 41.8% of anaerobic infections, *Clostridium* spp. (11.8%), *Prevotella*

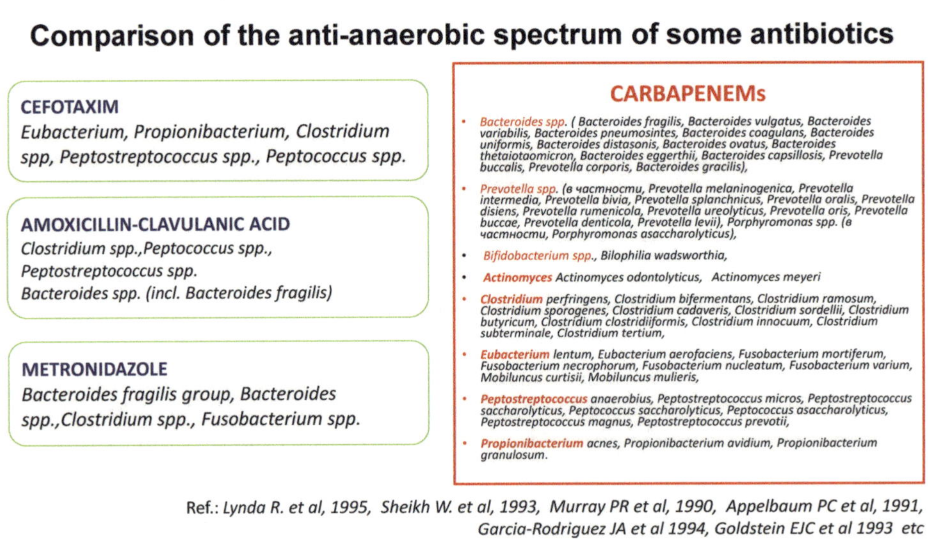

Figure 3.
Anti-anaerobic antibiotics vary in their coverage of anaerobes.

spp. (9.4%), *Peptostreptococcus* spp. (8.4%), etc. [35]. Usually, etiologically significant anaerobes are quite aggressive. Of course, in such patients, along with surgery, there are direct indications for the immediate use of anti-anaerobic antibiotics. At the same time, the duration of use of anti-anaerobic antibiotics should be reasonably limited, under the control of purulent focus sanitation, since prolonged use of anti-anaerobic antibiotics can cause irreversible disorders of the gut microbiota and lead to the development of sepsis.

It is important to take into account that anti-anaerobic antibiotics differ significantly in the "richness" of the spectrum of antimicrobial activity. **Figure 3** is based on data from the reference literature. It contains a list of the main genera/species of anaerobic bacteria that fall under the action of four antibiotics—for visual comparison. A priori, it can be assumed that when treated with carbapenems, the gut microbiota will experience the most negative effects. The consequence may be not only an increase in antibiotic resistance, but also an increase in the frequency of sepsis, which, unfortunately, is happening today in ICUs in different countries.

As a result of "fascination" with the excessive use of ultra-wide-spectrum antibiotics of the carbapenem group, in just the last 10–15 years, the world has found itself in a situation of total antibiotic resistance, which is now characteristic of the so-called hospital pathogens. Numerous facts of an increase in hospital deaths in recent years require little evidence, as evidenced by the titles of articles, for example, "Mortality Attributable to Bloodstream Infections Caused by Different Carbapenem-Resistant Gram-Negative Bacilli." Predictors of mortality in bloodstream infections caused by *Klebsiella pneumoniae*, etc. [36, 37].

The authors note that an increase in the number of carbapenem prescriptions has led to an increase in the number of carbapenem-resistant bacterial pathogens. The prevalence of extended-spectrum β-lactamase-producing and carbapenem-resistant Enterobacterales (CRE) has become a global public health problem. Colonization of patients' gut and spread in the surrounding hospital environment increases the risk of infection in patients and leads to poor disease prognosis. The review analyzes 25 studies in which attempts were made to apply various decolonization strategies to eliminate the presence of enterobacteria producing ESBL or resistant to carbapenems, but the results showed only a temporary and unstable effect [38].

The situation is most dangerous for the weakened categories of patients in the ICU, especially newborns. Due to the fact that in the first days of life, their gut microbiota has not yet been formed, the rapid spread of gram-negative bacteria resistant to carbapenems among newborns is concerning on a global scale. In the article Sisay et al. [39], 36 studies were included in the meta-analysis and systematic review. The study aimed to determine the combined prevalence of gram-negative bacteria resistant to carbapenem in African newborns who were suspected of having sepsis. The meta-analysis and systematic review included 36 studies. In total, among all gram-negative bacteria, resistance to imipenem was 35.57%, and to meropenem, 34.35%. The highest rates of resistance to carbapenems were in *Acinetobacter baumannii*—45.9%, and *Pseudomonas* spp.—43.0%. The authors point to the need to take urgent measures to regulate the strategy of control and use of antimicrobial therapy [39].

Another article also deals with neonatal sepsis, where Pseudomonas aeruginosa with multidrug resistance, expressing several carbapenemases, including NDM-1 and VIM-2, acts as an etiological agent. The authors express serious concern that there is simply nothing to treat such infections in the neonatal intensive care unit [40].

The situation in the adult ICU is no less alarming. The authors directly indicate in the titles of the articles the conclusion about the results of their research as an absolute fact, namely, "*in critically ill patients, anti-anaerobic antibiotics increase risk of adverse clinical outcomes*" [25]. Due to the progressive growth of carbapenem-resistant enterobacteria, especially carbapenemase-producing *Klebsiella pneumonia*, combination therapy with three or even four antibiotics is already being discussed [41, 42].

4.4 Increased sepsis and mortality

Large-scale studies conducted using systematic review and meta-analysis methods in Europe, North America, and Australia show high mortality rates for sepsis and shock septic [43]. There is every reason to assert that the increase in the incidence of sepsis and mortality is the result of ignoring the consequences of the destruction of the anaerobic gut microbiota by ultra-broad-spectrum antibiotics [37, 38, 44]. Global threat of the spread of carbapenem-resistant gram-negative bacteria has become uncontrollable, and doctors in ICU departments remain virtually unarmed [37, 38, 44].

Clinicians do not take into account the effect of such antibiotics on the patient's microbiota, and do not take into account the unforeseen consequences of such treatment. Numerous studies have shown that the destruction of anaerobic microorganisms in the gut using anti-anaerobic antibiotics affects systemic immunity and is associated with increased mortality in patients with sepsis. However, this knowledge has not yet been widely applied in clinical practice. A large retrospective cohort study involving nearly 16,000 emergency department patients showed that the widespread empirical use of anti-anaerobic antibiotics leads to a decrease in survival rates. It is important to note that there was a decrease in survival, including in the cohort of patients in whom infection was only suspected, but not confirmed [45]. In other words, some patients did not survive due to excessive stress on their microbiota with drugs that belong to the group of anti-anaerobic antibiotics.

The next study obtained results on changes in the composition of the rectal microbiota, which persisted in some recovered patients even a month after antibiotic treatment. These changes in the microbiota were associated with individual differences in reactions to cytokines, which allows us to build new ideas about the existence of mechanisms of "incomplete recovery." Thus, microbiota disorders during antibiotic therapy potentially contribute to the recurrence of infections and re-hospitalization after pneumonia. Unfortunately, this article does not provide information on which specific antibiotics were prescribed to patients hospitalized with pneumonia. Rectal microbiota are coupled with altered cytokine production capacity following community-acquired pneumonia hospitalization [46].

It is clear that measures for the prudent use of antimicrobials can lead to a minimization of unnecessary use of anti-anaerobic antibiotics, especially carbapenems, and will ensure the transition from carbapenems to alternative drugs. Despite this, doctors may be concerned that alternative treatments may have negative consequences, especially given that official treatment guidelines for patients do not yet contain such recommendations. In one of the recent reviews (2025) on carbapenem de-escalation as an antimicrobial stewardship strategy, 15 studies were selected that examined the replacement of carbapenems with alternative drugs that are not carbapenems. Almost all of these studies involved adult patients who were not in critical condition; in all cases, carbapenems were prescribed for the treatment of urinary tract infections, pneumonia, and skin and soft tissue infections. In 12 of the 15 studies, it was reported that the withdrawal of carbapenems and the transition to alternative drugs occurred 2–5 days after the start of treatment as part of the initiatives of the antibiotic stewardship program (ASP). Overall, carbapenem de-escalation was not associated with negative outcomes, higher rates of clinical failure, or mortality compared with other treatment-continuing groups [47]. However, it is unlikely that the data from this article can be useful to understand the future experience of applying microbiota conservation strategies.

In the context of the increasing resistance of hospital bacteria to carbapenems, pharmaceutical companies have proposed another method of "prolonging the life of carbapenems." A double-blind randomized clinical trial (31 ICU units in 4 countries) international randomized trial was organized and conducted, which included 607 patients with sepsis. The first group (n = 303) received meropenem continuous (3 g over 24 h), the second group (n = 304)—intermittent (1 g every 8 h) [48]. As expected, mortality in both groups did not differ; moreover, it was equally high (47% and 49%), meaning that patients who received meropenem in this study died more often from sepsis than from any other antibiotics.

5. Microbial metabolites and systemic impact

Discussing why treatment with anti-anaerobic antibiotics is associated with increased mortality, the authors of one study provide a forward-thinking commentary that "the increase in mortality is not entirely due to infections alone" [25] as if calling for a search for specific mechanisms related to the microbiota and its role in survival. Indeed, deep disturbances in the composition of the microbiota are manifested not only in the appearance of new inflammatory foci of infection and the development of sepsis caused by resistant bacteria. The suppression of obligate anaerobes and a sharp decrease in biodiversity lead to a disruption of microbiota metabolism. This phenomenon has not been studied much yet, but disorders of the metabolic pathways associated with the microbiota are inevitable in critically ill patients, and even more so under the pressure of certain antibiotics.

Along with the growing understanding of the importance of the microbiota, questions naturally arise about how bacteria living in the human gut can interact with a macroorganism and send signals of well-being or distress. An assumption was made about the existence of low molecular weight metabolites of microbial origin, which are the normal end products of the metabolic "conveyor" of the microbiota entering the blood from the gut. If the microbiota is damaged or suppressed, that is, it represents a "pathobiota" [6]. The metabolic pathways cannot be completed, and intermediate metabolic products enter the bloodstream, which, upon entering the systemic circulation, can have an adverse effect on the organs and systems of the host body (**Figure 4**). This hypothesis about small molecules originating from microbes (SMOM) was first published in an article by Beloborodova and Osipov in 2000 [49].

The hypothesis about SMOM is that the inflammatory process is triggered by disorders at the low molecular weight level involving small (signaling) microbial molecules. It follows that the development of a complex multifactorial process called sepsis is associated with a deep decompensation of SMOM homeostasis. This hypothesis is logically justified by understanding the expediency of creating a system of signaling molecules in ontogenesis that would serve as intermediaries for the exchange of information between the microbiota and host cells. Various types of fatty acids (hydroxy acids, branched, unsaturated, and cyclopropane acids), aldehydes, alcohols, and phenylcarboxylic substances, which are rare in healthy people but increase several times in severe diseases, have been detected using gas chromatography-mass spectrometry.

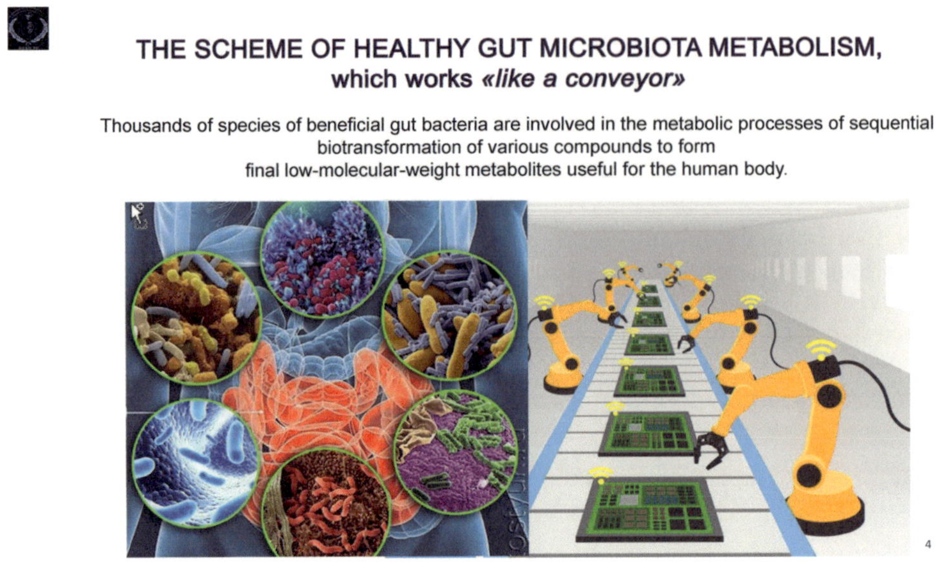

Figure 4.
The conveyor of microbial metabolism in the gut normally supplies the host's body with beneficial small molecules—microbial metabolites.

Further studies have shown that microbial metabolites of aromatic structure, so-called aromatic microbial metabolites (AMM), have the greatest diagnostic and pathogenetic significance for assessing the state of the microbiota and the degree of its involvement in the development of the septic process. The three most clinically significant aromatic metabolites of tyrosine and phenylalanine, such as 4-hydroxy-phenyl lactic acid (HPhLA), 4-hydroxyphenylacetic acid (HPhAA), and phenyl lactic acid (PhAA), have been conventionally referred to as "sepsis-associated metabolites" (**Figure 5**) [50].

The metabolic pathways leading to the accumulation or removal of AMM are closely interrelated in the human body with the metabolism of the gut microbiota. A decrease in biological diversity, including during treatment with anti-anaerobic antibiotics, is manifested by a violation of the metabolic functions of the microbiota, which leads to an excess of AMM in the blood. Our experimental studies *in vitro* have confirmed the ability of human microbiota bacteria to produce and consume the above-mentioned clinically significant microbial substances.

AMM can be used to assess the severity of patients' condition and the risk of death in critically ill patients, to predict complications in cardiac and abdominal surgery, neurosurgery, etc., as well as to monitor the effect of treatment in ICUs. An article published in the journal Shock describes the mechanism of participation of aromatic microbial metabolites in the development of septic shock (**Figure 6**) [51].

To date, methods with GC-MS and HPLC-MS equipment have been validated and used for clinical research to accurately measure the concentration of AMM in the blood. Based on a large volume of clinical and laboratory data, models have been developed to predict an unfavorable outcome for patients admitted to the intensive care unit in critical condition with various inflammatory and infectious foci, using methods of modern mathematical analysis. It has been proven that the concentration of eight microbial metabolites circulating in the blood can predict the risk of mortality for a particular patient. The ROC analysis confirmed that the reliability

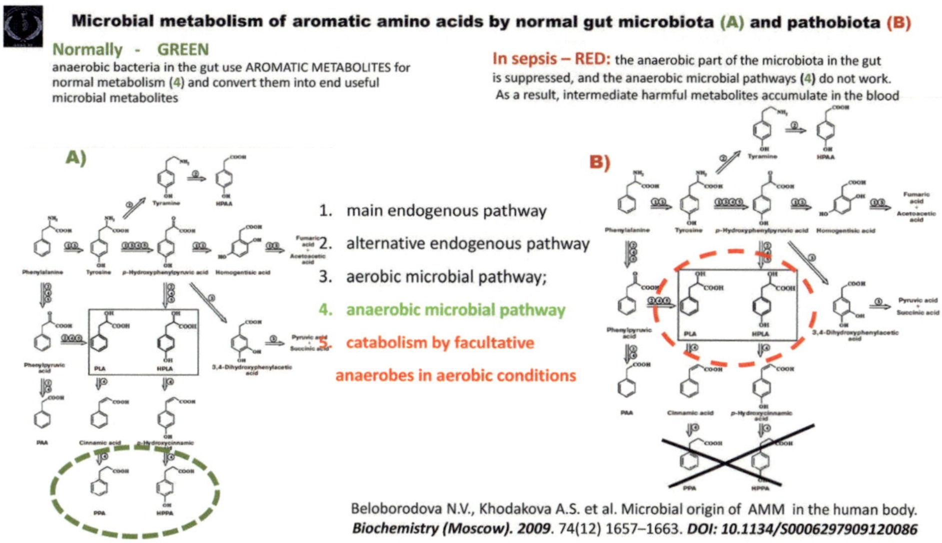

Figure 5.
The end products of microbial biotransformation of aromatic amino acids are useful, while the intermediate ones are harmful.

of the prognosis for microbial metabolites is not inferior to the generally accepted international multiparametric scales (e.g., APACHE II), which indirectly indicates the enormous pathogenetic significance of the dysfunction of the microbiota and the products of its pathological metabolism in the development of the systemic inflammatory process [52]. A clinical study has confirmed the important role of initial and acquired disorders of microbiota metabolism in the development of postoperative complications in patients with aortic aneurysm by monitoring the level of aromatic metabolites circulating in the blood before surgery and in the early postoperative period [53] Detailed data on aromatic microbial metabolites are presented in one of the latest books published by Springer on modern biomarkers for intensive care [54].

On the other hand, in a "healthy" microbiota, metabolites with a different chemical structure may be important for triggering the chain of immunological mechanisms of an adequate immune system response to the introduction of an infectious agent. For example, indole propionic acid (IPA), a metabolite of the aromatic amino acid tryptophan, has the property of "turning on" macrophages by activating aryl hydrocarbon receptor (AHR) [55]. Other positive clinical effects were also identified, for example, associated with IPA in the blood of patients when the postoperative period proceeded without signs of delirium [56].

Among the large group of short-chain fatty acids SCFA, microbial metabolites of the gut microbiota, butyrate has been of particular interest in recent years. Butyrate is produced by butyrogenic bacteria of the large intestine and, along with other SCFA, participates in maintaining the integrity of the intestinal barrier. The interest in this metabolite is not accidental, since the destruction of the intestinal barrier is an important mechanism in the development of sepsis. The translocation of bacteria from the gut in critically ill patients is associated, among other things, with the destruction of the normal microbiota by anti-anaerobic antibiotics. In addition, *Clostridium difficile* infection and nosocomial diarrhea are also associated with a deficiency of butyrate-producing bacteria [57].

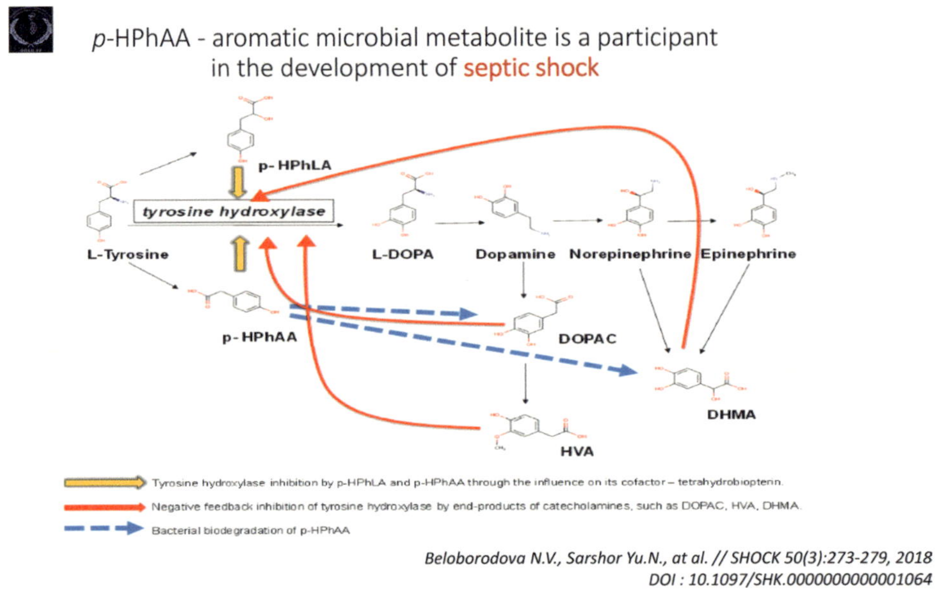

Figure 6.
The products of microbial metabolism of aromatic amino acids may be involved in the genesis of septic shock.

An interesting paper Kullberg et al. [45], presents the results from two observational, population-based microbiome studies. A more favorable course of infections and a reduced risk of hospitalization were noted in those patients whose microbiota had a sufficiently high relative content of butyrate-producing bacteria. In this study, the 16S rRNA gene sequencing method was used to evaluate the gut microbiota. The composition of the microbiota was analyzed in a fairly large number (more than 10,500 participants) of the study. The authors found that those participants whose microbiota was rich in butyrate-producing bacteria had a significantly lower risk of developing infections. On the contrary, the number of infections, hospitalizations, and even deaths was significantly higher in the group with the altered microbiota (which does not produce butyrate). The results were verified and proved to be comparable in two independent cohorts of two European countries, the Netherlands and Finland, as part of the HELIUS study. One of the results of this study is an important suggestion that modulation of the human microbiota can reduce the risk of severe infections [58].

6. Clinical implications and future directions

Under conditions of suppression of the anaerobic microbiota in the gut of patients, a non-competitive environment is created for the appearance and accumulation of "problematic" multiplicity-resistant strains of gram-negative enterobacteria (primarily *Klebsiella pneumonia*), non-fermenting bacteria (*Pseudomonas* spp. and *Acinetobacter* spp.), as well as gram-positive methicillin-resistant staphylococci (MRS) and vancomycin-resistant enterococci (VRE).

In recent years, several retrospective analyses have been published comparing the benefits and harms of two different treatment strategies for hospitalized patients using a relatively narrow range of antibiotics that do not suppress anaerobic bacteria, or broader-spectrum antibiotics with anti-anaerobic activity. For short, the first group is called "non-anaerobic" and the second "anaerobic." Thus, the expected side effect of antibiotics of the first group on the obligate anaerobic gut microbiota is neutral (preservation), and antibiotics of the second group are negative (suppression). By dividing patients into groups depending on treatment and comparing the results in a retrospective analysis, the authors identify certain anti-anaerobic antibiotics that are more often used in patients of a particular ICU. The lists of anti-anaerobic antibiotics in clinical trials by different authors may vary slightly, as shown in **Table 1**. These are the results of randomized cohort studies conducted in hospitalized patients from different countries.

The negative impact on the results of treatment with anti-anaerobic antibiotics was found in randomized cohort studies conducted among hospitalized patients in different years. It is logical to assume that it would be beneficial to limit or avoid the empirical use of the anti-anaerobic antibiotics listed in this table in order to preserve the microbiota.

Clinical studies are emerging that even partial restrictions on the use of anti-anaerobic antibiotics give positive results. For example, the study involved 34 patients in the neurological intensive care unit, who were treated with either "non-anaerobic" narrow-spectrum antibiotics or broad-spectrum antibiotics meropenem or piperacillin/tazobactam upon admission according to clinical indications. To assess the microbiota dynamics, a rectal smear was taken from each patient before the start of antibiotic therapy and 5–7 days after it. The study found that most antibiotic resistance genes were significantly more likely to occur after treatment with broad-spectrum drugs (meropenem or piperacillin/tazobactam) [61].

Of great interest is the article by Chanderraj et al. [25], which revealed a link between treatment with anti-anaerobic antibiotics and adverse clinical outcomes in 3032 patients with mechanical ventilation. After analyzing the gut microbiota of 116 patients and conducting animal experiments, the authors found evidence of a causal effect of anti-anaerobic antibiotics on infectious and non-infectious outcomes [25].

Antibiotics play a key role in the treatment of ICU patients with pneumonia, intra-abdominal infection, postoperative complications and sepsis. When prescribing empirical antibiotic therapy, doctors often try to be safe by choosing an antibiotic or a

"Anti-anaerobic" parenteral antibiotics	Author	The consequences	Ref.
Ampicillin, carbenicilline, clindamycine, linkomycine	Van der Waaij [59, 60]	Colonization of the gut	**Figure 1**
Meropenem piperacillin/ tazobactam	Nielsen et al. [61]	Growth of antibiotic resistance genes	doi:10.3390/ microorganisms9122542
Piperacillin-tazobactam, metronidazole, ampicilline-sulbactam, ceftriaxone, meropenem	Chanderraj et al. [25]	Decreased ventilator-associated pneumonia (VAP)-free survival	doi:10.1183/13993003.00910-2022
Piperacillin-tazobactam, meropenem, metronidazole, clindamycin, amoxicillin—clavulanic acid	Kullberg [45]	Decrease in survival of critically ill patients	doi:10.1183/13993003.00413-2023

Table 1.
The use of anti-anaerobic antibiotics was associated with negative consequences of the treatment in randomized cohort clinical studies.

combination of 2 or 3 drugs with a wider range of antibacterial activity than is really necessary in a given clinical situation. The authors of a large study are convinced that anaerobes are rarely the causative agents of sepsis, at the same time, anti-anaerobic antibiotics are used very often in the ICU, of course, "for the best of reasons" or "just in case."

The authors of one of the latest studies on this topic, Chanderraj et al. managed to implement the design of a comparative study in which large groups of sepsis patients receiving combined antibiotic therapy differed fundamentally based on the use or non-use of an anti-anaerobic antibiotic [62]. This study included 7569 patients with sepsis, compared the treatment results in two groups that met the study criteria and were comparable in key indicators. In the first group, patients (n = 4523) received vancomycin and piperacillin-tazobactam, and in the second group (n = 3046)—vancomycin and cefepime. The results of the instrumental analysis showed that the use of piperacillin and tazobactam was associated with an absolute 5.0% increase in mortality within 90 days, an increase in the duration of organ failure, mechanical ventilation, and the use of vasopressors compared to cefepime. Based on the data obtained, the authors draw an important reasoned conclusion that anti-anaerobic antibiotics (in this case, piperac) should not be prescribed to patients with sepsis unless there are clear indications for anti-anaerobic therapy.

7. Conclusion

Global statistics show that the problem of sepsis remains extremely relevant, and, despite significant efforts, the mortality rate from sepsis remains high. In this regard, interest in scientific research on the human gut microbiota has increased significantly in recent years. A strategy known since the 1980s as "selective decontamination of the digestive tract" (SDD) experienced a "rebirth" in the 2000s. Enteral administration of nonabsorbable antibiotics, whose spectrum of action is aimed at suppressing only aerobic bacteria in the gut, allows for the preservation of gut anaerobes. The positive clinical effect of SDD has been proven in numerous clinical studies. It is noteworthy that positive results in clinical studies of SDD are achieved by eliminating bacterial competition in the gut in favor of obligate anaerobes, that is, by "restoring" the functioning of the human microbiome. SDD is particularly effective for preventive purposes in high-risk patients, as it helps to reduce the infectious complications and risk of sepsis.

Patients admitted to the clinic with an infection (pneumonia, peritonitis, abdominal infections, high-risk of sepsis, etc.) always require parenteral administration of antibiotics. Recent studies have shown that certain antibiotics (e.g., carbapenems or modern penicillin-based drugs, including those combined with beta-lactamase inhibitors), when administered for empirical therapy intravenously/intramuscularly, can eliminate anaerobic bacteria in the gut of hospitalized patients. Due to depletion of intestinal anaerobic bacteria, the mucous membranes of the digestive tract may be colonized with hospital-acquired pathogens with multiple drug resistance, which can lead to decreased immunoreactivity, increased risk of bacteria entering the bloodstream, and an increased risk of secondary infectious complications and mortality. Large-scale clinical observation studies demonstrate that the use of anti-anaerobic antibiotics significantly increases the risk of developing sepsis.

Published results on the benefits of limiting/avoiding the use of anti-anaerobic antibiotics to preserve the microbiota in hospitalized patients deserve serious attention. However, it should be noted that at the time of writing, there are no randomized controlled studies available to confirm the validity of the results. The results of retrospective studies have their limitations and cannot serve as a basis for direct clinical recommendations. For example, a retrospective study compares the outcomes of treatment in two groups, one of which includes all patients who received anti-anaerobic antibiotics. However, it cannot be ruled out that anti-anaerobic antibiotics (such as carbapenems) were used more frequently in the most severe patients in life-threatening emergencies, and the initial severity of these patients' conditions may have led to more adverse outcomes and influenced the results of the retrospective analysis.

There is no doubt that randomized controlled trials are needed in the near future to help select optimal antimicrobial therapy strategies that take into account the impact on the gut microbiota to reduce the incidence of sepsis and improve patients' survival.

- "Evidence suggests that the widespread use of empirical anti-anaerobic antibiotics in sepsis may be harmful" Chanderraj et al. [62].
- "Empirical antibiotic therapy for sepsis: save the anaerobic microbiota" Kullberg et al. [63].

These two phrases from scientific articles published in 2024–2025 could serve as an epigraph for this chapter. In summary, the review presented in this chapter reflects promising trends in antimicrobial therapy strategies that take into account the important role of microbiota, and today they should be considered as a basis for developing protocols and conducting prospective clinical trials.

Author details

Natalia Beloborodova
Federal Research and Clinical Center of Intensive Care Medicine and Rehabilitation, Moscow, Russian Federation

*Address all correspondence to: nvbeloborodova@yandex.ru

References

[1] Singer M et al. The sepsis definitions task force the third international consensus definitions for sepsis and septic shock (Sepsis-3). JAMA. 2016;**315**(8):801-810. DOI: 10.1001/jama.2016.0287

[2] Ojima M, Motooka D, Shimizu K, et al. Metagenomic analysis reveals dynamic changes of whole gut microbiota in the acute phase of intensive care unit patients. Digestive Diseases and Sciences. 2016;**61**(6):1628-1634. DOI: 10.1007/s10620-015-4011-3

[3] Lankelma JM, van Vught LA, Belzer C, et al. Critically ill patients demonstrate large interpersonal variation in intestinal microbiota dysregulation: A pilot study. Intensive Care Medicine. 2016;**43**:59-68. DOI: 10.1007/s00134-016-4613-z

[4] Jacobs MC, Haak BW, Hugenholtz F, et al. Gut microbiota and host defense in critical illness. Current Opinion in Critical Care. 2017;**23**(4):257-263. DOI: 10.1097/MCC.0000000000000424

[5] McDonald D, Ackermann G, Khailova L, et al. Extreme dysbiosis of the microbiome in critical illness. mSphere. 2016;**1**(4):e00199-16. DOI: 10.1128/mSphere.00199-16

[6] Zaborin A, Smith D, Garfield K, et al. Membership and behavior of ultra-low-diversity pathogen communities present in the gut of humans during prolonged critical illness. MBio. 2014;**5**(5):e01361-14. DOI: 10.1128/mBio.01361-14

[7] Chernevskaya E, Klimenko N, Pautova A, et al. Host-microbiome interactions mediated by phenolic metabolites in chronically critically ill patients. Metabolites. 2021;**11**(2):122. DOI: 10.3390/metabo11020122. Available from: https://www.mdpi.com/2218-1989/11/2/122

[8] Haak BW, Levi M, Wiersinga WJ. Microbiota-targeted therapies on the intensive care unit. Current Opinion in Critical Care. 2017;**23**(2):167-174. DOI: 10.1097/MCC.0000000000000389

[9] Price R, MacLennan G, Glen J, on behalf of the SuDDICU Collaboration. Selective digestive or oropharyngeal decontamination and topical oropharyngeal chlorhexidine for prevention of death in general intensive care: Systematic review and network meta-analysis. BMJ. 2014;**348**:g2197. DOI: 10.1136/bmj.g2197

[10] Johanson WG Jr, Pierce AK, Sanford JP, et al. Nosocomial respiratory infections with gram-negative bacilli. The significance of colonization of the respiratory tract. Annals of Internal Medicine. 1972;**77**:701-706. DOI: 10.7326/0003-4819-77-5-701

[11] van Saene HKF, Damjanovic V, Murray AE, et al. How to classify infections in intensive care units—The carrier state, a criterion whose time has come? The Journal of Hospital Infection. 1996;**33**:1-12. DOI: 10.1016/s0195-6701(96)90025-0

[12] Avni T, Levcovich A, Ad-El DD, at al. Prophylactic antibiotics for burns patients: Systematic review and meta-analysis. BMJ. 2010;**340**:c241. DOI: 10.1136/bmj.c241

[13] Barajas-Nava LA, López-Alcalde J, Roqué i Figuls M, et al. Antibiotic prophylaxis for preventing burn wound infection. Cochrane Database of

Systematic Reviews. 2013;**6**:CD008738. DOI: 10.1002/14651858.cd008738.pub2

[14] Stoutenbeek CP, van Saene HK, Miranda DR, et al. The effect of selective decontamination of the digestive tract on colonisation and infection rate. Intensive Care Medicine. 1984;**10**:185-192

[15] de Jonge E, Schultz MJ, Spanjaard L, et al. Effects of selective decontamination of digestive tract on mortality and acquisition of resistant bacteria in intensive care: A randomised controlled trial. Lancet. 2003;**362**(9389):1011-1016. DOI: 10.1016/s0140-6736(03)14409-1

[16] Silvestri L, de la Cal MA, van Saene HK. Selective decontamination of the digestive tract: The mechanism of action is control of gut overgrowth. Intensive Care Medicine. 2012;**38**(11):1738-1750. DOI: 10.1007/s00134-012-2690-1

[17] Roquilly A, Marret E, Abraham E, Asehnoune K. Pneumonia prevention to decrease mortality: A systematic review and meta-analysis. Clinical Infectious Diseases. 2015;**60**(1):64-75. DOI: 10.1093/cid/ciu740

[18] Wittekamp BHJ, Oostdijk EAN, Cuthbertson BH, Brun-Buisson C, Bonten MJM. Selective decontamination of the digestive tract (SDD) in critically ill patients: A narrative review. Intensive Care Medicine. 2020;**46**(2):343-349. DOI: 10.1007/s00134-019-05883-9

[19] Abis GSA, Stockmann HBAC, Bonjer HJ, van Veenendaal N, van Doorn-Schepens MLM, Budding AE, et al. Randomized clinical trial of selective decontamination of the digestive tract in elective colorectal cancer surgery (SELECT trial). The British Journal of Surgery. 2019;**106**(4):355-363. DOI: 10.1002/bjs.11117

[20] Roos D, Dijksman LM, Tijssen JG, et al. Systematic review of perioperative selective decontamination of the digestive tract in elective gastrointestinal surgery. The British Journal of Surgery. 2013;**100**(12):1579-1588. DOI: 10.1002/bjs.9254

[21] Waked K, Schepens M. State-of-the-art review on the renal and visceral protection during open thoracoabdominal aortic aneurysm repair. Journal of Visceral Surgery. 2018;**4**:31. DOI: 10.21037/jovs.2018.01.12

[22] Safdar N, Said A, Lucey MR. The role of selective digestive decontamination for reducing infection in patients undergoing liver transplantation: A systematic review and meta-analysis. Liver Transplantation. 2004;**10**(7):817-827. DOI: 10.1002/lt.20108

[23] De Waele JJ, Leroux-Roels I, Depuydt P. Selective digestive decontamination—Pro. Intensive Care Medicine. 2023;**49**:979-981. DOI: 10.1007/s00134-023-07100-0

[24] Xue L, Ding Y, Qin Q, Liu L, et al. Assessment of the impact of intravenous antibiotics treatment on gut microbiota in patients: Clinical data from pre- and post-cardiac surgery. Frontiers in Cellular and Infection Microbiology. 2023;**12**:1043971. DOI: 10.3389/fcimb.2022.1043971

[25] Chanderraj R, Baker JM, Kay SG, Brown CA, Hinkle KJ, Fergle DJ, et al. In critically ill patients, anti-anaerobic antibiotics increase risk of adverse clinical outcomes. The European Respiratory Journal. 2023;**61**(2):2200910. DOI: 10.1183/13993003.00910-2022

[26] Moore WE, Holdeman LV. Human fecal flora: The normal flora of 20 Japanese-Hawaiians. Applied

Microbiology. 1974;**27**(5):961-979. DOI: 10.1128/am.27.5.961-979.1974

[27] Lagier JC, Armougom F, Million M, et al. Microbial culturomics: Paradigm shift in the human gut microbiome study. Clinical Microbiology and Infection. 2012;**18**(12):1185-1193. DOI: 10.1111/1469-0691.12023

[28] Wan X, Yang Q, Wang X, Bai Y, Liu Z. Isolation and cultivation of human gut microorganisms: A review. Microorganisms. 2023;**11**(4):1080. DOI: 10.3390/microorganisms11041080

[29] Tierney BT, Yang Z, Luber JM, Beaudin M, Wibowo MC, Baek C, et al. The landscape of genetic content in the gut and oral human microbiome. Cell Host & Microbe. 2019;**26**(2):283-295.e8. DOI: 10.1016/j.chom.2019.07.008

[30] Brook I, Wexler HM, Goldstein EJ. Antianaerobic antimicrobials: Spectrum and susceptibility testing. Clinical Microbiology Reviews. 2013;**26**:526-546. DOI: 10.1128/CMR.00086-12

[31] Kioka MJ, DiGiovine B, Rezik M, Jennings JH. Anaerobic antibiotic usage for pneumonia in the medical intensive care unit. Respirology. 2017;**22**(8):1656-1661. DOI: 10.1111/resp.13111

[32] Marin-Corral J, Pascual-Guardia S, Amati F, et al. Aspiration risk factors, microbiology, and empiric antibiotics for patients hospitalized with community-acquired pneumonia. Chest. 2021;**159**(1):58-72. DOI: 10.1016/j.chest.2020.06.079

[33] Bergan T, Nord CE, Thorsteinsson SB. Effect of meropenem on the intestinal microflora. European Journal of Clinical Microbiology & Infectious Diseases. 1991;**10**(6):524-527. DOI: 10.1007/BF01963943

[34] Watabe J, Mizutani J, Suzuki H, et al. Short communication: Effect of a new carbapenem antibacterial agent, meropenem, on normal human faecal microflora. Microbial Ecology in Health and Disease. 1992;**5**(6):315-321. DOI: 10.3109/08910609209141553

[35] Park Y, Choi JY, Yong D, Lee K, Kim JM. Clinical features and prognostic factors of anaerobic infections: A 7-year retrospective study. The Korean Journal of Internal Medicine. 2009;**24**(1):13-18. DOI: 10.3904/kjim.2009.24.1.13

[36] Tumbarello M, Pierluigi Viale P, Claudio VC, et al. Predictors of mortality in bloodstream infections caused by carbapenemase–producing *K. pneumoniae*: Importance of combination therapy. Clinical Infectious Diseases. 2012;**55**(7):943-950. DOI: 10.1093/cid/cis588

[37] Falcone M, Tiseo G, Carbonara S, et al. Mortality attributable to bloodstream infections caused by different carbapenem-resistant gram-negative Bacilli: Results from a Nationwide study in Italy (ALARICO network). Clinical Infectious Diseases. 2023;**76**(12):2059-2069. DOI: 10.1093/cid/ciad100

[38] Zhang HJ, Wang HW, Tian FY, et al. Decolonization strategies for: A systematic review and meta-analysis. Science Reports. 2024;**14**(1):24349. DOI: 10.1038/s41598-024-75791-6

[39] Sisay A, Asmare Z, Kumie G, et al. Prevalence of carbapenem-resistant gram-negative bacteria among neonates suspected for sepsis in Africa: A systematic review and meta-analysis. BMC Infectious Diseases. 2024;**24**(1):838. DOI: 10.1186/s12879-024-09747-6

[40] Daaboul D, Osman M, Kassem II, Yassine I, Girlich D, Proust A, et al.

Neonatal sepsis due to NDM-1 and VIM-2 co-producing *Pseudomonas aeruginosa* in Morocco. The Journal of Antimicrobial Chemotherapy. 2024;**79**(7):1614-1618. DOI: 10.1093/jac/dkae153

[41] Jacobs DM, Safir MC, Huang D, Minhaj F, Parker A, Rao GG. Triple combination antibiotic therapy for carbapenemase-producing *Klebsiella pneumoniae*: A systematic review. Annals of Clinical Microbiology and Antimicrobials. 2017;**16**(1):76. DOI: 10.1186/s12941-017-0249-2

[42] Falagas ME, Lourida P, Poulikakos P, Rafailidis PI, Tansarli GS. Antibiotic treatment of infections due to carbapenem-resistant Enterobacteriaceae: Systematic evaluation of the available evidence. Antimicrobial Agents and Chemotherapy. 2014;**58**(2):654-663. DOI: 10.1128/AAC.01222-13

[43] Bauer M, Gerlach H, Vogelmann T, et al. Mortality in sepsis and septic shock in Europe, North America and Australia between 2009 and 2019—Results from a systematic review and meta-analysis. Critical Care. 2020;**24**:239. DOI: 10.1186/s13054-020-02950-2

[44] Jean SS, Harnod D, Hsueh PR. Global threat of carbapenem-resistant gram-negative bacteria. Frontiers in Cellular and Infection Microbiology. 2022;**12**:823684. DOI: 10.3389/fcimb.2022.823684

[45] Kullberg RFJ, Schinkel M, Wiersinga WJ. Empiric anti-anaerobic antibiotics are associated with adverse clinical outcomes in emergency department patients. The European Respiratory Journal. 2023;**61**:2300413. DOI: 10.1183/13993003.00413-2023

[46] Kullberg RFJ, Brands X, Klarenbeek AM, et al. Rectal microbiota are coupled with altered cytokine production capacity following community-acquired pneumonia hospitalization. iScience. 2022;**25**(8):104740. DOI: 10.1016/j.isci.2022.104740

[47] Gardner A, Nieberg P, Sakoulas G, et al. Carbapenem de-escalation as an antimicrobial stewardship strategy: A narrative review. JAC-Antimicrobial Resistance. 2025;**7**(2):dlaf022. DOI: 10.1093/jacamr/dlaf022

[48] Monti G, Bradić N, Marzaroli M, et al. Continuous vs intermittent meropenem administration in critically ill patients with sepsis: The MERCY randomized clinical trial. Journal of the American Medical Association. 2023;**330**(2):141-151. DOI: 10.1001/jama.2023.10598

[49] Beloborodova NV, Osipov GA. Small molecules originating from microbes (SMOM) and their role in microbes–host relationship. Microbial Ecology in Health and Disease. 2000;**12**(1):12-21. DOI: 10.1080/089106000435545

[50] Beloborodova NV, Khodakova AS, Bairamov IT, Olenin AY. Microbial origin of phenylcarboxylic acids in the human body. Biochemistry (Mosc). 2009;**74**(12):1350-1355. DOI: 10.1134/s0006297909120086

[51] Beloborodova NV, Sarshor YN, Bedova AY, Chernevskaya EA, Pautova AK. Involvement of aromatic metabolites in the pathogenesis of septic shock. Shock. 2018;**50**(3):273-279. DOI: 10.1097/SHK.0000000000001064

[52] Pautova AK, Samokhin AS, Beloborodova NV, Revelsky AI. Multivariate prognostic model for predicting the outcome of critically ill patients using the aromatic metabolites detected by gas chromatography-mass

spectrometry. Molecules. 2022;**27**(15):4784. DOI: 10.3390/molecules27154784

[53] Beloborodova N, Pautova A, Grekova M, Yadgarov M, Grin O, Eremenko A, et al. Microbiota metabolism failure as a risk factor for postoperative complications after aortic prosthetics. Biomedicine. 2023;**11**(5):1335. DOI: 10.3390/biomedicines11051335

[54] Beloborodova NV. Serum aromatic microbial metabolites as biological markers in intensive care. In: Rajendram R, Preedy VR, Patel VB, editors. Biomarkers in Trauma, Injury and Critical Care. Biomarkers in Disease: Methods, Discoveries and Applications. Cham. Chapter 13: Springer; 2023. pp. 245-268. DOI: 10.1007/978-3-031-07395-3_64

[55] Keskey RC, Xiao J, Hyoju S, et al. Enterobactin inhibits microbiota-dependent activation of AhR to promote bacterial sepsis in mice. Nature Microbiology. 2025;**10**:388-404. DOI: 10.1038/s41564-024-01882-9

[56] Zhou X, Wu X, Wu Y, et al. Indole-3-propionic acid, a gut microbiota metabolite, protects against the development of postoperative delirium. Annals of Surgery. 2023;**278**(6):e1164-e1174. DOI: 10.1097/SLA.0000000000005886

[57] Singh V, Lee G, Son H, et al. Butyrate producers, "The Sentinel of Gut": Their intestinal significance with and beyond butyrate, and prospective use as microbial therapeutics. Frontiers in Microbiology. 2023;**13**:1103836. DOI: 10.3389/fmicb.2022.1103836

[58] Kullberg RFJ, Wikki I, Haak BW, et al. Association between butyrate-producing gut bacteria and the risk of infectious disease hospitalisation: Results from two observational, population-based microbiome studies. Lancet Microbe. 2024;**5**(9):100864. DOI: 10.1016/S2666-5247(24)00079-X

[59] Van der Waaij D, Berghuis-de Vries JM. Selective decontamination of enterobacteriaceae species from the digestive tract in mice and monkeys. The Journal of Hygiene. 1974;**72**:205-211

[60] Van der Waaij D. History of recognition and measurement of colonization resistance of the digestive tract as an introduction to selective gastrointestinal decontamination. Epidemiology and Infection. 1992;**109**(3):315-326

[61] Nielsen KL, Olsen MH, Pallejá A, Ebdrup SR, Sørensen N, Lukjancenko O, et al. Microbiome compositions and resistome levels after antibiotic treatment of critically ill patients: An observational cohort study. Microorganisms. 2021;**9**(12):2542. DOI: 10.3390/microorganisms9122542

[62] Chanderraj R, Admon AJ, He Y, et al. Mortality of patients with sepsis administered piperacillin-tazobactam vs cefepime. JAMA Internal Medicine. 2024;**184**(7):769-777. DOI: 10.1001/jamainternmed.2024.0581

[63] Kullberg RFJ, Haak BW, Chanderraj R, Prescott HC, Dickson RP, Wiersinga WJ. Empirical antibiotic therapy for sepsis: Save the anaerobic microbiota. The Lancet Respiratory Medicine. 2025;**13**(1):92-100. DOI: 10.1016/S2213-2600(24)00257-1

Chapter 2

Enhancing Survival Following High-Risk Surgery by Modulating Actionable Items within the Pathobiome

Olga Zaborina, Alexander Zaborin, Robert Keskey and John Alverdy

Abstract

Infection-related complications are common among surgical patients and result in morbidity and mortality. Despite the use of antibiotics, infection-related complications continue to occur. While the main sources of postoperative infections have been assumed to be due to some types of perioperative external contamination, an increasing number of studies have suggested an endogenous source when intestinal pathogens disseminate to cause the syndrome of "gut-origin sepsis." In this context, the composition and function of the intestinal microbiome appear to play a crucial role. Patients are at increased risk for these complications when the microbiome shifts to a pathobiome and becomes predominated by pathogenic organisms instead of the usual beneficial organisms. It is important to note that the pathobiome may manifest its most aggressive form when conditions are such that the host experiences significant physiological stress (i.e., surgical stress) which often involves long periods of inanition, surgical injury and antibiotic exposure. In this review, pathobiomes will be defined and compared to microbiomes and incorporate the mechanisms by which virulence activation of opportunistic pathogens within pathobiomes emerge over the course of surgical injury. We will define the physiologic context associated with surgical stress as well the suppression of host defense mechanisms that can occur by abnormally functioning pathobiomes. Along with this line of reasoning, there are several possible strategies for preventing postoperative infection-related complications by improving the composition and function of the intestinal microbiota and suppressing the virulence of the pathobiome thereby mitigating the impact of surgical stress factors on the activation of pathogen virulence expression.

Keywords: microbiome, pathobiome, post-operative infection, surgical stress signals, nutrient scarcity, FMT, DietPreHab, Pi-PEG

1. Introduction

Postoperative infection-related complications continue to occur despite mandated antisepsis protocols, including skin decontamination, changing of gloves during surgery, environmental sterility and the liberal use of prophylactic antibiotics. Postoperative infection related complications include the development of wound infections, deep organ-space infections, anastomotic leaks, and systemic infections leading to organ failure and death (i.e., pneumonias, bacteremias, sepsis). Although it is presumed that most infections originate from contamination by an environmental pathogen, emerging evidence suggests that the patient's own endogenous microbiome could be the site of origin of many postoperative infections that arise from the stress of the surgery itself, the use of antibiotics that select for pathogenic microbes and long periods of poor or artificial nutrition.

In order to enhance the field of surgery and improve patient care, we should set the bar at achieving zero postoperative infections which will require us to go beyond the utilization of antibiotics and sterilization.

This chapter outlines the various predisposing factors leading to a pathogenic microbiome, its negative role during surgical stress, and the actionable items that may mitigate this process.

2. Definition of pathobiomes *versus* microbiomes

2.1 The main characteristics of healthy intestinal microbiomes

2.1.1 Domination of Firmicutes and Bacteroidetes and high diversity

A healthy intestinal microbiome is characterized by a high degree of microbial diversity and consists of various species and genera. While, on the phylum level, Firmicutes, Bacteroidetes, Proteobacteria, and Actinobacteria are the most prevalent phyla, there are two primary dominating phyla, Firmicutes and Bacteroidetes, which encompass numerous diverse genera of anaerobic bacteria [1–3]. Most of the diversity and species belonging to these phyla reside in the colon as it has a higher bacterial density and greater diversity as compared to the small intestine [4, 5].

2.1.2 Stability/resilience of the microbiome to recover from a perturbation

Another important characteristic of a healthy intestinal microbiome is their resilience and stability. This is reflected by the state of dynamic equilibrium that exists between beneficial bacteria and their capacity to compete with opportunistic pathogens. Furthermore, healthy microbiomes often can recover and return to a stable state after various disturbances, including dietary changes, infections, and antibiotic use (i.e., resilience) [6, 7]. For example, it has been shown that individuals on a low-fiber diet can experience a decline in the diversity of their gut microbiota [8]. However, when they switch to a fiber-rich diet, studies indicate that the microbiome can recover its diversity within days, resulting in an increase in beneficial bacterial groups such as Firmicutes and Bacteroidetes that ferment dietary fiber into short-chain fatty acids (SCFAs), which are necessary for gut health [9].

Post-Antibiotic Recovery: Antibiotic exposure is common and increasing worldwide and has repeatedly been shown to negatively impact the gut microbiome [10]. After a course of antibiotics, such as amoxicillin or clindamycin, the gut microbiome often experiences a significant shift in its composition and function [11, 12]. In many cases, a healthy individual's microbiome can start to recover to its initial state within weeks to months after antibiotic cessation [13, 14]. Beneficial bacteria such as *Lactobacillus* and *Bifidobacterium* may gradually re-establish their populations as microbiome diversity increases and the overall core community stabilizes [13, 15]. Although some probiotics have been demonstrated to reduce some of the side effects of antibiotics (diarrhea), their ability to restore the gut microbiota is limited [16]. The best agent to restore the microbiota has been the provision of an autologous FMT (obtained from preAbx exposure). This latter observation assumes that the refaunation actually occurs from the native communities of bacteria specific to that individual person. Therefore, from an ecological perspective, in some cases, simply providing competing species or communities may actually be detrimental [16]. It may be for this reason that many probiotic formulations, when rigorously tested, cannot live up to their purported claims.

Infectious Recovery: During an infectious stress, the gut microbiome can diminish in diversity and experience a temporary increase in pathogen predominance [17, 18]. However, once the infection resolves, a healthy host returns to a stable microbiome state within weeks, re-establishing beneficial bacteria like *Faecalibacterium prausnitzii* that produce SCFAs and contribute to gut health [19]. Probiotics may help restore the microbiome to a more stable and healthy state and accelerate recovery in microbial diversity and function [20–23]. These examples highlight the resilience of healthy microbiomes and their ability to recover to stable states after various disturbances, including dietary changes, infections, and antibiotic use. Many factors are in play, including the original state of microbial diversity, the extent to which beneficial bacteria are present and the host's overall dietary and environmental lifestyle.

2.1.3 Production of healthy metabolites, that is, short-chain fatty acids and indoles

As a reflection of its compositional state, the functional capacity of the microbiome is its most important characteristic and encompasses a wide range of metabolic functions including, but not limited to, fermentation of dietary fiber, synthesis of vitamins, and metabolism of amino acids, etc. [1]. The microbiome breaks down complex carbohydrates that are not digestible by the human gastrointestinal tract, commonly referred to as microbiota accessible carbohydrates [24]. This fermentation process leads to the production of SCFAs such as acetate, propionate, and butyrate, which not only provide energy to colon cells but also contribute to overall gut health [25, 26]. SCFAs also maintain the integrity of the intestinal barrier and prevent excessive inflammatory responses [27–29].

Certain gut bacteria synthesize essential vitamins, including vitamin K and several B vitamins (B12, B2, and folate) [30–32], which are crucial cofactors for various metabolic functions in the body and, in some cases, have been tied to SCFA production by the gut microbiota [33–36].

The metabolism of amino acids is a vital aspect of gut function and overall health [37, 38]. Amino acids serve as building blocks for proteins, precursors for neurotransmitters and other biomolecules, and have various roles in metabolic pathways [39, 40]. Certain amino acid metabolites are primarily or exclusively produced by gut microbes. Key amino acid-derived metabolites that are exclusively generated by microbial activity include indoles, metabolites derived from tryptophan by various gut bacteria, particularly from the Firmicutes and Bacteroidetes phyla, the indole-3-propionic acid (IPA), indole-3-acetic acid (IAA), and indole-3-aldehyde (IAld) [41–44]. Indoles are the signaling molecules that play a crucial role by binding to and activating the aryl hydrocarbon receptor (AhR), an intracellular host cell receptor that can serve as a transcription factor that mediates the effects of environmental and dietary ligands [42, 45, 46]. The binding of gut microbiota-derived metabolites to AhR initiates a signaling cascade that results in the translocation of the receptor to the nucleus, where it influences gene expression [47, 48]. The activation of AhR by indole derivatives, for example, modulates immune responses such as a differentiation of regulatory T cells (Tregs) and production of anti-inflammatory cytokines [48, 49]. It also activates the transition of macrophages from pro-inflammatory M1-like phenotype to anti-inflammatory M2-like phenotype with increased phagocytic activity [42]. The interaction between indoles and AhR promotes the growth of beneficial microorganisms and suppresses harmful bacteria, supporting a balanced microbiome [49, 50]. Indoles are known for their potential roles in cancer prevention [51]. Through AhR activation, they can induce the expression of genes involved in detoxification and apoptosis, providing protective effects against cancer development [52]. Indole signaling through AhR may have implications for neurodevelopment and neurological diseases [53, 54]. Finally, indoles possess antimicrobial properties and can inhibit the growth of various pathogens, potentially contributing to the host's defense mechanisms [50, 55, 56].

2.2 Key characteristics of pathobiomes

2.2.1 Domination by opportunistic pathogens and reduced diversity of microbiota

Many factors can contribute to the disruption of a healthy microbiome and its transition to a pathobiome and are associated with the life-history of the host. These include exposure to: antibiotics, high-fat/high-sugar/low-fiber diets, acute infections, psychological and physiological stressors, modern lifestyle reduced contact with soil and animals, advanced age, exposure to medications (i.e., proton pump inhibitors (PPIs), non-steroidal anti-inflammatory drugs (NSAIDs), corticosteroids), chronic diseases such as obesity, diabetes, inflammatory bowel diseases (IBD), major surgery, a lack of physical activity, smoking, and excessive alcohol consumption [57–69].

The transition from a healthy microbiome to a pathobiome is characterized by specific changes in the composition and diversity of microbial communities that may be associated with certain diseases and severity of illness [57, 70]. Here we will outline how common medical co-morbidities and diseases are associated with changes in the gut microbiota. In IBD such as Crohn's disease and ulcerative colitis, there is an association of the disease state with a decrease in bacterial diversity [71, 72]. Specific groups, such as Firmicutes (notably the family *Ruminococcaceae*), may be decreased, while the abundance of pathogenic bacteria such as *Escherichia coli* and other certain members of the Proteobacteria phylum increase [73–75]. Obesity is a common medical comorbidity in the western world and has been associated with a higher ratio of Firmicutes to Bacteroidetes in the gut microbiota [76]. One popular hypothesis is that this increased ratio is associated with a higher capacity for energy harvest from the diet due to increased Firmicutes [77]. This indicates that certain microbiota may extract more calories from the same amount of food, thereby contributing to excess energy consumption and potential weight gain [78]. Type 2 diabetes has also been linked to changes in the gut microbiome, including increased levels of *E. coli*, and decreased levels of beneficial bacteria, including some *Lactobacillus* species [79]. An altered gut microbiome may contribute to insulin resistance and inflammation [80]. In cardiovascular diseases, an increased abundance of Trimethylamine-producing bacteria (i.e., *Escherichia, Prevotella*) is associated with elevated levels of trimethylamine-N-oxide (TMAO), which is linked to atherosclerosis [81]. In conditions such as depression and anxiety, research suggests an alteration in gut microbiota composition, including reduced diversity and an increase in inflammatory bacteria. For instance, certain strains of *Prevotella* may be elevated, which could influence gut-brain communication [82, 83]. Individuals with allergic diseases or asthma may exhibit a reduced diversity in their gut microbiomes [84]. The presence of beneficial microbes (e.g., *Bifidobacteria* and *Lactobacilli*) may be decreased, while bacteria associated with inflammation increase [85]. However, association does not necessarily translate to causation and further work is needed in this field.

While reduced diversity is a common feature of a pathobiome, ultra-low diversity has been observed in the gut microbiota of critically ill patients [86]. An increase of opportunistic pathogens at the expense of beneficial microbes, reflecting a shift in the balance of the microbial community, sometimes yields to the domination of only one to three species of bacterial pathogens such as *Enterococcus faecalis*, *Klebsiella pneumoniae*, *Enterobacter* sp., etc. [86–88]. These pathogenic communities often include *Candida* spp. [86]. To conclude, reduced diversity and a predominance of opportunistic pathogens are among the many features of a pathobiome [89]. The extent to which this observation is causative to poor outcome following critical illness remains to be confirmed.

2.2.2 Corresponding genotype and phenotype analyses

Culture-based microbiological analysis of stool samples collected from critically ill patients often corresponds to their genetic analysis. For example, in an analysis of stool in critically ill patients, a predominance of *Enterococcus* determined by 16S rRNA gene sequencing corresponded to *E. faecalis* determined by culture analysis, or in the case of another patient's stool, *Enterobacteriaceae* detected by genetic analysis corresponded to its presence via culture analysis [86]. This degree of concordance between genetic and culture analysis has not been previously observed in healthy microbiomes [86].

2.2.3 Failure of pathobiomes to return to their initial state following a perturbation

The stability of pathobiomes refers to the ability of a pathogenic microbial community to maintain its composition and functionality in the context of disease [90]. Pathobiomes often display a certain degree of stability, allowing them to resist disturbances and maintain their pathogenic characteristics. This stability can be due to the presence of specific microbial taxa that thrive in the dysbiotic environment and possess mechanisms that enable them to predominate over beneficial microbes. The interactions among microorganisms within a pathobiome can contribute to its stability [90]. For example, pathogenic microbes may produce metabolites that inhibit the growth of competing beneficial bacteria, thereby reinforcing their own dominance and stability in the community [91]. Pathobiomes are often associated with chronic inflammation, which can create an environment conducive to the stability, and thus, retention of pathogenic microbes [92]. Specific environmental conditions are needed to establish certain pathobiomes and thus support their stability [57, 67]. These conditions include altered levels of oxygen, decreased health promoting metabolites (i.e., SCFAs), and the absence of healthy bacteria that typically outcompete for resources.

A change of environmental factors can impact the stability of pathobiomes during which time they may become challenged to recover to a baseline stable state after various disturbances. For instance, it has shown that certain pathobiomes do not return to their initial state after major surgery [93]. NOTE: major surgery involves a general anesthetic, exposure to antibiotics, a major operative intervention, postoperative pain and the use of opioids, disordered sleep, and a significant period of inanition [94]. In comparison, a normal baseline gut microbiome has been observed to return to its original composition after 7 days of a major surgical intervention [93]. This highlights that changes in the environmental factors affecting the pathobiome—such as antibiotic exposure, dietary changes, physiological stress—do not always lead to a return to its initial state. Instead, such changes may result in a new stable state of equilibrium among pathobiomes that can persist life-long and may have later health implications.

The above observations may explain, in part, the occasional inefficiency and unpredictability of fecal microbial transplants (FMT) to restore the microbiome [95, 96]. An FMT might lose its capacity to propagate and dominate when entering an environment that favors emergence and stability of a pathobiome. While such "replacement therapies" might be conceptually attractive, understanding the durable ecological impact of one's overall life-history may make simple replacement of a "healthy" microbiome challenging.

Several examples demonstrate how new pathogenic communities may be durably embedded in one's gut ecosystem after antibiotic treatments or diet changes. In patients with *Clostridium difficile* infection (CDI), the initial pathobiome characterized by high levels of *Clostridium difficile* can change significantly after antibiotic treatment [97]. However, several studies show that the pathobiome—dominated by pathogenic *C. difficile*—often does not revert to its original state. Instead, a new composition may stabilize, with *C. difficile* persisting along with reduced diversity and abundance of protective commensal bacteria that could prevent recurrence, thus leading to a higher risk of further infection [98]. Antibiotic treatment may be intended to remove pathogenic bacteria but may unintentionally select for other pathogens, like opportunistic bacteria (e.g., *E. faecalis*), leading to a new stable pathobiome profile that defends the presence and persistence of the ongoing dysbiosis and inflammation without full return of the original infecting pathogen [86]. Aggressive antibiotic therapy in critically ill patients directed at remove pathogens detected in blood and other organs has been shown to replace one pathogenic community (e.g., *E. faecium* + *C. albicans*) with another (*E. faecalis* + *Serratia marcescens* + *Klebsiella oxytoca* + *C. albicans)*; strains that have proven to be aggressive and virulent [86].

In patients with IBD, the initial pathobiome might be characterized by specific inflammatory microbes such as certain *E. coli* strains [99]. After treatment (such as immunosuppressants or biological therapies), the pathobiome does not necessarily revert to its original state, retaining pathogenic characteristics or harboring new inflammatory species that contribute to chronic inflammation and ongoing disease activity. In the context of obesity, a pathobiome may include specific strains of *Firmicutes* that contribute to energy harvest [77]. Following weight loss interventions, such as dietary restriction or bariatric surgery, the pathobiome does not fully revert to its pre-obesity composition [100, 101]. Instead, new persister states may form and can be characterized by different pathogenic traits or by a persistent imbalance associated with a pro-inflammatory condition, thus demonstrating the importance of antecedent patho-environmental conditions that allow for the predominance of pathogenic strains [102].

2.2.4 Shift from beneficial metabolic processes to those that may promote disease

The metabolic output of pathobiomes may be characterized by a shift from beneficial metabolic processes to those that may promote disease [103]. Here are some key characteristics of the metabolic patterns of pathobiomes:

1. Decreased production of SCFAs. In healthy microbiomes, beneficial bacteria ferment dietary fibers to produce SCFAs like acetate, propionate, and butyrate, which play important roles in maintaining gut health and regulating immune function [104]. When pathobiomes persist, the production of SCFAs may be reduced due to decreased populations of fiber-fermenting bacteria, leading to increased gut permeability and inflammation [105]. An acute decrease of SCFAs in stool has been observed immediately after trauma or surgical stress [106–110]. The use of antibiotics, particularly broad-spectrum antibiotics, which significantly disrupts the gut microbiota, can lead to a decrease in SCFA-producing bacteria such as *F. prausnitzii* and *Roseburia* and a consequent decrease of SCFAs production [111]. Psychological stress, such as that caused by anxiety or depression, can alter the gut microbiota in both composition and function [112, 113]. This situation can lead to reduced levels of SCFAs. Acute inflammation, such as following infection or a bout of IBD can lead to a decrease in SCFA levels [114]. For instance, during an episode of acute colitis, the diversity of SCFA-producing bacteria may decline, leading to reduced production of SCFAs like butyrate; sudden changes in diet such as a rapid shift away from fiber-rich foods to a low-fiber or high-fat diet intended to lessen the inflammatory process [115]. When dietary fibers are eliminated, in addition to the obvious loss of the main source of SCFAs, the microbial populations that ferment these fibers tend to decline, leading to an overall decrease in SCFAs. SCFAs play critical roles in maintaining intestinal barrier integrity, regulating inflammation, and influencing energy metabolism [27, 116]. As such, their precipitous reduction can have significant health implications.

2. Indoles are primarily produced by specific gut microbes that utilize fermentation pathways to metabolize the dietary substrate tryptophan [43, 117–119]. Microbial species produce different tryptophan catabolites such as via conversion to tryptamine, indolelactic acid, and indolepropionic acid by *Clostridium* spp., [120, 121], and to indoleacrylic acid and indolepropionic acid by *Peptostreptococcus* spp. [122]. Furthermore, *Lactobacillus* spp. metabolize tryptophan to indolealdehyde and indolelactic acid [118, 123], and Ruminococcus converts it to tryptamine by a tryptophan decarboxylase [120]. Several *Bacteroides* species produce indolelactic acid and indoleacetic acid, and *Bifidobacterium* spp. produce indolelactic acid [124]. Indole formation occurs via the action of the enzyme tryptophanase (TnaA), which is expressed in many Gram-negative, as well as Gram-positive bacterial species, including *E. coli*, *Clostridium* spp. and *Bacteroides* spp. [118]. The production of indoles from the amino acid tryptophan by various gut bacteria typically occurs

in environments where oxygen levels are low, such as in the intestines [125]. Even *Bacteroides*, certain strains of *Clostridium*, and *Lactobacillus* mainly contribute to indole production, there are some representatives of Proteobacteria, including *E. coli*, *Serratia marcescens*, *Enterobacter aerogenes*, *Proteus vulgaris*, *Salmonella enterica*, *K. pneumoniae* that can produce indole derivative indole-3 acetic acid from tryptophan through various pathways, including the indole-3-acetamide pathway, the indole-3-pyruvic acid pathway, the indole-3-acetonitrile pathway, the tryptamine pathway, and the tryptophan side-chain oxidase pathway [126]. In Proteobacteria, bacteria producing indoe-3-acetic acid (I3AA), are mainly plant-associated since I3AA belongs to the family of auxin indole derivatives [126]. Some bacteria can use an oxidative pathway of tryptophan metabolism through the action of bacterial tryptophan 2,3-dioxygenase (TDO) to produce kynurenine instead [127, 128]. Although this pathway is mainly attributed to mammal cells, predicted bacterial tryptophan 2,3-dioxygenases (InterPro IRP017485) were confirmed by experimental characterizations and by conserved operon structure for many bacterial members (https://www.ebi.ac.uk/interpro/entry/InterPro/IPR017485/).

The kynurenine pathway is also implicated for its immunosuppressive functions. Abnormal or inappropriate activation of this pathway can lead to accumulation of kynurenine—a metabolite associated with various disease states [129]. Elevated levels of kynurenine and its metabolites have been implicated in psychiatric disorders such as depression and schizophrenia [130–132]. Increased tryptophan catabolism through the kynurenine pathway, particularly under inflammatory conditions, may influence mood-regulating pathways and contribute to depressive symptoms [133]. Kynurenine is associated with neuroinflammation and neurodegeneration [134]. In conditions like Alzheimer's disease and Parkinson's disease, altered kynurenine metabolism may exacerbate neuroinflammatory processes and lead to neurotoxic effects, contributing to the progression of these diseases [135]. Kynurenine levels can be increased in response to inflammatory cytokines (e.g., interferon-gamma) during chronic inflammatory diseases [136]. Additionally, an imbalance in kynurenine metabolism can promote chronic inflammation and has been linked to conditions such as rheumatoid arthritis [137] and IBD [138]. Kynurenine is produced in higher quantities in some types of cancer. It is often overactive in cancer and linked to tumor growth and metastasis [139–141]. Tumors can exploit the kynurenine pathway to suppress immune responses, enabling tumor growth and survival [142]. Elevated levels of kynurenine have been correlated with poorer prognosis in certain cancers. Kynurenine has been associated with cardiovascular diseases, as increased kynurenine levels may correlate with endothelial dysfunction and systemic inflammation, risk factors for atherosclerosis and other cardiovascular complications [143]. In summary, kynurenine is considered a metabolite associated with disease due to its involvement in various pathological processes, including neuroinflammation, immune modulation, and metabolic dysregulation.

Given that the intestinal environment, especially the colon, is in general a site in which there is little to no oxygen, the involvement of the aerobic kynurenine pathway of tryptophan metabolism by bacterial species is questionable. However, researchers have recently demonstrated that physiological stress, such as occurs following major surgery (i.e., a partial hepatectomy in the mouse), can lead to an elevation of the oxygen concentration at the upper crypt levels in cecum— a site in that is typically characterized by strict anaerobic conditions [144]. Yet, these circumstances can be associated with compositional changes in the microbiota when the upper crypt becomes colonized with facultative anaerobes, including *Proteobacteria* species accompanied by an acute decrease of anaerobic bacteria [144]. As such, elevated oxygen in such sites may result in a change in the composition and function of the microbiota [125]. It is possible, therefore, that under such environmental circumstances (i.e., in which microenvironmental zones of elevated oxygen exist), a shift in microbial tryptophan metabolism from indoles to kynurenine may occur.

It is important to note that pathogens can actively inhibit AhR activation through the production of AhR-antagonists (i.e., enterobactin) [42]. From another site, pathogens like uropathogenic *E. coli* are able to induce kynurenine production in mammalian cells [145]. Interestingly, both indoles and kynurenine are AhR ligands [146, 147], but the context in which they are produced may dictate whether they are associated with health *versus* disease. When indole activates AhR, it generally promotes beneficial pathways associated with gut health, anti-inflammatory effects, and support of host defenses [42, 148, 149]. This is in contrast to the immunosuppressive and potentially harmful effects associated with increased kynurenine production.

The domination of opportunistic pathogens in pathobiomes may lead to the accumulation of a variety of toxins that contribute to their pathogenic potential, causing direct cellular damage, disrupted host immune responses, or systemic sequelae. As an example, an imbalance of gut microbiota after antibiotic treatment may allow *C. difficile* to overgrow and produce its primary toxins: toxin A (enterotoxin) and toxin B (cytotoxin). These toxins disrupt intestinal epithelial function and activate inflammation, causing diarrhea and colitis [150, 151].

Staphylococcus aureus can produce various toxins, including Alpha-toxin, a pore-forming toxin that damages cell membranes and leads to cell lysis [152], enterotoxins that cause food poisoning and can act as superantigens, leading to systemic inflammatory responses [153], Toxic Shock Syndrome Toxin (TSST-1) associated with toxic shock syndrome and can lead to severe systemic effects [154, 155].

Certain pathogenic strains of *E. coli* (i.e., enterohemorrhagic *E. coli* (EHEC) produce Shiga toxin: that can cause severe gastrointestinal disease and may lead to hemolytic uremic syndrome [156]. Enterotoxigenic *E. coli* (ETEC) produce Heat-Labile and Heat-Stable Toxins, causes of diarrhea [157].

Pseudomonas aeruginosa produces several toxins including Exotoxin A, which inhibits protein synthesis in host cells by ADP-ribosylation, contributing to tissue damage and inflammation [158]; Elastase, which breaks down elastin and other tissue components (i.e., collagens) facilitating tissue invasion and promoting inflammation [159, 160]; pyocyanin, which is a potent redox-active compound capable of generating reactive oxygen species (ROS) within host cells. This leads to oxidative stress, damaging cellular components such as lipids, proteins, and DNA, which can impair cell function and promote cell death [161]. By interfering with mitochondrial respiration, pyocyanin can disrupt ATP production in host cells. This inhibition can reduce energy availability for essential cellular processes, contributing to cell dysfunction and death, impair phagocytic activity, reduce chemotactic responses, and modulate cytokine production, diminishing the host's ability to combat infections, induce apoptosis in various cell types, including respiratory epithelial cells [162]. The induction of programmed cell death can contribute to tissue damage and facilitate bacterial colonization and infection [163]. Similarly, certain strains of *E. faecalis* can produce cytolysin, a toxin that can cause hemolysis and damage host tissues, potentially contributing to infections [164, 165]. Based on the above conclusions, the characteristics of a microbiome versus a pathobiome are displayed in **Table 1**.

	Microbiome	Pathobiome
Microbiota composition at the Phylum level	Dominance of Firmicutes and Bacteroidetes	Dominance of one type of Phylum and/or Significant dominance of Proteobacteria
Microbiota composition at the Genus level	High level of diversity, no significant dominance of any one genus	Dominance of one or more genera, usually opportunistic pathogens, often resistant to multiple antibiotics
Culture analysis	Microbiological cultures do not correspond to genetic analysis	Opportunistic pathogens are cultured according to the genetic analysis
Impact on outcome after surgery in model experiments	Supports the barrier function of the intestinal epithelium, preventing systemic infection, and stabilizes the immune system	Reduces the barrier function of the intestinal epithelium, leads to systemic infection, imbalances the immune system, high rate of postoperative mortality
The metabolic pattern	Beneficial metabolic processes to host and entire microbiota	The metabolic pattern is characterized by a shift from beneficial metabolic processes to those that may promote disease.
Toxins accumulation	No toxins accumulation	The accumulation of variety of toxins that contribute to pathogenic potential of pathobiomes, causing direct cellular damage, disrupt host immune responses, and/or lead to systemic effects.

Table 1.
Characteristics of a microbiome versus *a pathobiome.*

3. Pathogens within pathobiomes activate a program of virulence in response to surgical stress signals

Domination of opportunistic pathogens within a pathobiome imposes a risk for increasing their pathogenicity as opportunistic pathogens can activate their virulence in response to surgical stress signals via the release of host immune elements, cytokines, physico-chemical cues within the environment, microbiome signals and direct regulatory mechanisms with the host-pathogen interface.

3.1 Changes in hormone, opioids and cytokine levels

Aseptic surgical trauma can induce a state of inflammation via the release of inflammatory cytokines, opioids and hormones [166–168]. These host stress-derived signals can be "sensed" by opportunistic pathogens, following which they "respond" accordingly. Responses can include activating their own quorum sensing signaling molecules, which leads to multiple downstream virulence circuits. As a reminder, quorum sensing (QS) is a regulatory mechanism that allows bacteria to communicate with each other in a density-dependent manner [169]. This process enables bacterial populations to coordinate their behavior and gene expression based on their cell density, facilitating assemblage behavior, including the expression of various virulence genes.

3.1.1 Interferon-γ as a signal for activating the virulence system in Pseudomonas aeruginosa

During surgical injury, the body experiences physical stress and tissue injury resulting in activation of the immune and inflammatory system. This stress can trigger the release of pro-inflammatory cytokines, including IFN-γ [170]. IFN-γ can bind to OprF receptor on *P.* aeruginosa (**Figure 1**) [171]. OprF (Outer membrane protein F) is a porin protein, it forms channels in the outer membrane that allow the passage of small molecules and ions [175]. OprF has been shown to interact with C4-HSL [176]. The presence of OprF on the outer membrane helps facilitate the binding and uptake of C4-HSL [176]. When IFN-γ binds to OprF, the accumulation of C4-HSL reaches the threshold concentration required for activation of the quorum sensing machinery to occur at a lower cell density [171]. It is possible that binding of IFN-γ modifies OprF, thus increasing its interaction with C4-HSL. Accordingly, all pathways regulated by C4-HSL's interaction with RhlR lead to the production of virulence factors, including pyocyanin and the outer membrane adhesion the PA-I lectin, all of which are induced by IFN-γ at an earlier cell density [171].

3.1.2 Opioids as a signal for activating the virulence of Pseudomonas aeruginosa

The gastrointestinal tract is one of the most common sources of endogenous opioids in the body due to the presence of the complex enteric nervous system, often termed the "second brain" [177]. During surgical stress, opioids (beta-endorphin, dynorphin, enkephalins) are released into the bloodstream in high concentrations and can have multiple effects on the physiology of the gastrointestinal tract and immune function [178].

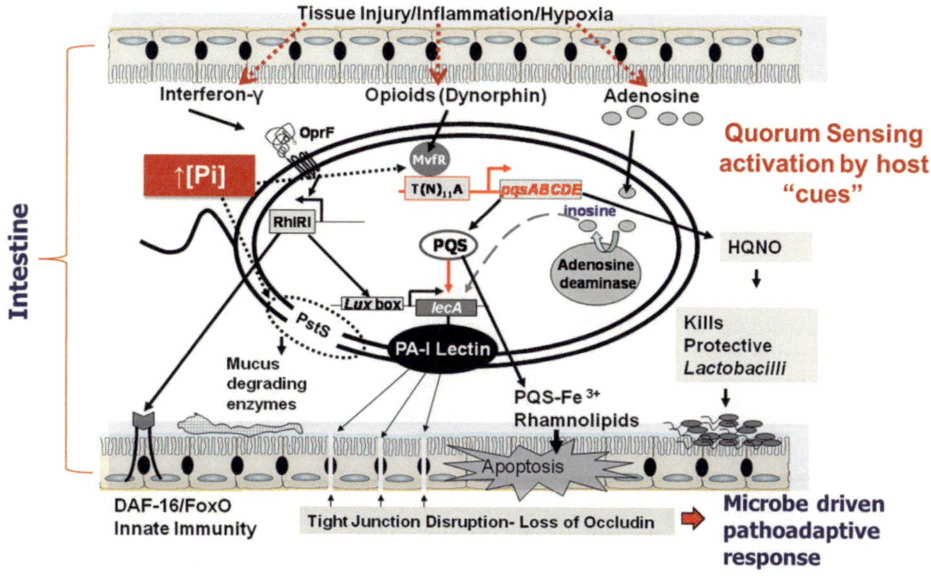

Figure 1.
Summarized effect of host stress derived signals on virulence activation in P. aeruginosa. *This figure is created based on the findings described in Refs. Wu et al. [171], Zaborina et al. [172], Patel et al. [173], and Zaborin et al. [174]. During tissue injury/inflammation the released IFN-gamma binds to* P. aeruginosa *receptor protein OprF that leads to the activation of RhlRI pathway of quorum sensing system followed by the expression of virulence factors such as PA-I lectin that disrupts the tight junctions (i.e., loss of occludin) in epithelial cells and modulating innate immunity via DAF-16/FoxO pathway. Opioid dynorphin through MvfR-dependent Pseudomonas quinolone signaling activation induces overproduction of PQS and 2-heptyl-4-hydroxyquinoline N-oxide (HQNO) leading to increased production of PA-I lectin and rhamnolipids causing apoptosis of epithelial cells, while HQNO is involved in the killing of intestine epithelium-protective* Lactobacilli. *Adenosine release during hypoxia to protect intestinal epithelium is converted to inosine by* P. aeruginosa *adenosine deaminase. Phosphate limitation in environment of gut in stressed host induces a global virulence in P.* aeruginosa *via PstS-PhoB system and activation of MvfR-PQS pathway of quorum sensing.*

Opioids are primarily known for their role in pain modulation and their effects on the central nervous system; however, their potential influence on bacterial quorum sensing (QS) has been an area of interest in research. Opioids can be classified based on their affinity for the three primary types of opioid receptors: mu (μ), kappa (κ), and delta (δ) receptors [179]. Among the different classes of opioids, the greatest virulence response by the opportunistic pathogen *P. aeruginosa* has been observed with exposure to kappa-opioids, including an endogenous opioid peptide dynorphin and synthetic ligand U-50,488 [172]. Dynorphin is primarily synthesized and released by neurons of the enteric nervous system (ENS), which governs the function of the gastrointestinal tract [180]. The release of dynorphin in response to inflammatory signals such as cytokines or immune activation is an important aspect of its physiological role [181]. During conditions like IBD or infection, inflammatory mediators can lead to increased dynorphin release, which influences pain perception and other gut processes [182]. The mechanism of activation of quorum sensing in *P. aeruginosa* by dynorphin and U-50,488 includes MvfR-PQS pathway. PQS triggers the synthesis of many virulence factors; PA-I lectin, which increases the permeability of the cellular epithelium; cyanide hydroxide; pyocyanin, which suppresses the host immune system; biofilm; etc. [172]. The precursor of PQS, 2-heptyl-4-quinolone, actively suppresses the growth of probiotic bacteria, helping to expand the population niche of *P. aeruginosa* in the intestine (**Figure 1**) [172].

A less pronounced but still significant effect on pyocyanin production by *P. aeruginosa* has been demonstrated for morphine [183]. Morphine is primarily known as a selective ligand for the mu-opioid receptor (MOR), but it also has effects on other opioid receptors, including the kappa-opioid receptor (KOR), although its affinity for KOR is significantly lower than for MOR [184]. The mechanism of activation was not elucidated in details; however, it was demonstrated that morphine attracts *P. aeruginosa* that creates high density local concentrations of bacteria at morphine spots with increased production of virulent factor PA-I lectin [183].

3.1.3 Stress hormones as a signal for activating the virulence system

Surgical stress triggers the release of stress hormones such as cortisol and catecholamines (e.g., epinephrine and norepinephrine), which can both modulate the immune response as well as enhance the virulence of opportunistic pathogens [185]. Norepinephrine is part of the stress response in the host, and bacteria can "sense" stress-related signals from the host and "respond" accordingly [186]. This sense-and-respond circuitry allows pathogens to adapt their strategies to exploit the host environment effectively. Some bacteria possess receptors that can sense catecholamines like norepinephrine [187]. These adrenergic receptors can initiate intracellular signaling pathways that promote virulence factor expression. For example, *P. aeruginosa* has been shown to have a receptor RetS that binds norepinephrine, leading to enhanced expression of virulence genes [188, 189]. In response to norepinephrine, certain bacteria may enhance the production of QS signals such as acyl-homoserine lactones (AHLs), leading to the expression of virulence factors such as toxins, adhesins, and enzymes that promote bacterial survival and pathogenicity [190–193]. In *S. aureus*, for instance, norepinephrine can stimulate the production of toxins [194].

3.1.4 Intestinal hypoxia is recognized by pathogens as an inducing virulence signal

Hypoxia, which occurs when cellular oxygen demand exceeds supply, is a common feature associated with bacterial infection [195–197]. Pathogens can recognize and respond to extracellular end products of intestinal hypoxia that are released after activation of HIF-1-alpha [173, 198]. It has been shown that the ability of *P. aeruginosa* to metabolize adenosine to inosine may represent a subversive microbial virulence strategy that deprives the epithelium of the cytoprotective actions of adenosine (**Figure 1**) [173].

4. The effect of nutrient availability on the virulence of pathobiomes

Surgical stress changes nutrient availability to microbial communities resulting in the rise of pathogenic bacteria capable of accessing intracellular nutrients for growth and survival [199, 200]. The release of physico-chemical "cues" such as when a scarcity of extracellular nutrients such as iron and phosphate occur in gut, itself can enhance the virulence of certain bacteria [201–203].

4.1 Iron limitation during surgical stress

Iron limitation in the gut during surgical stress is an important physiological response that can affect both the host and microbial communities [204]. Iron is an essential nutrient required for various biological processes, including hemoglobin formation, cellular respiration, and DNA synthesis [205]. However, its availability is tightly regulated in the body to prevent oxidative damage to organs and cells as well as to limit the pathogenesis of infections [206, 207]. The gut is the primary site for iron absorption, and normal iron status relies on dietary intake and homeostatic mechanisms that control absorption and storage [208]. Surgical stress induces an acute inflammatory response characterized by the release of cytokines and other immune mediators. This inflammatory response can lead to changes in iron metabolism [209]. During inflammation, the liver produces hepcidin, a hormone that regulates iron homeostasis [210]. Hepcidin inhibits iron absorption from the gut and sequesters iron in macrophages and hepatocytes, thereby reducing iron availability in the bloodstream and elsewhere [211]. This can lead to functional iron deficiency even if total body iron stores are adequate. As hepcidin levels increase in response to surgical stress and inflammation, intestinal absorption of iron decreases due to reduced expression of the iron transporter ferroportin, a transmembrane protein that acts as a channel for iron export from cells, particularly in the intestine, liver, and macrophages [212]. It is the only known iron exporter, facilitating the transport of ferrous iron (Fe^{2+}) across the cell membrane and plays a critical role in iron transport from inside cells to the extracellular space. Ferroportin's activity is regulated by the hormone hepcidin, which is produced by the liver in response to iron concentrations and the extent of inflammation [212]. When hepcidin binds to ferroportin, it triggers internalization and degradation of the protein, thus reducing iron export from cells [212]. Reduced expression of ferroportin leads to lower availability of dietary iron in the gut. Iron limitation can have a profound influence on the gut microbiome, given that many, if not most, bacteria require iron for growth and metabolism.

Other key iron-related proteins, such as transferrin and ferritin, also play significant roles in iron transport and storage, while lactoferrin and soluble transferrin receptor serve as additional components in the complex regulation of iron metabolism [213]. During surgical stress, the regulation of transferrin, ferritin, lactoferrin, and the soluble transferrin receptor are influenced by the acute inflammatory response, changes in iron metabolism, and iron demand.

Pathogenic bacteria adapt to low-iron conditions through the expression of siderophores—secreted molecules that bind iron more effectively than host proteins—allowing them to sequester iron from the host and enhance their own growth, survival and virulence [214, 215]. Pathogens such as *P. aeruginosa* can secrete pyoverdine and pyochelin [216], members of the Enterobacteriaceae such as *K. pneumoniae*, *E. coli*, *S. enterica*, *Enterobacter cloacae*, *S. marcescens*, *Proteus mirabilis*, *Citrobacter freundii* as well as *Strepromyces* spp. secret a common siderophore enterobactin [217], *S. aureus* produces staphyloferrin A and B [218], *Vibrio cholerae* – vibriobactin [219], *Bacillus subtilis* – bacillibactin [220], *B. anthracis* – bacillibactin and petrobactin [221], *Mycobacterium tuberculosis* – mycobactin [222], and *S. marcescens* – serratiochelin [223]. The ability to produce these various siderophores reflects the sense and response circuitry of bacteria as they adapt to varying ecological niches. Under normal conditions, bacteria maintain intracellular iron levels through iron acquisition mechanisms, including uptake through permeases and transporters [224]. When environmental iron is limited, bacteria "sense" this deficiency and "respond" by increasing siderophore production. In many bacteria, including *E. coli* and *P. aeruginosa*, the Fur protein acts as a repressor of siderophore production when iron is present in sufficient quantities [225]. When iron is depleted, Fur undergoes a conformational change, releasing its repression, and allowing siderophore biosynthesis genes to be expressed. Certain Gram-positive bacteria may use alternative systems or proteins like iron-dependent transcription factors that respond directly to iron availability [226]. Once released, siderophores bind tightly to ferric ions, forming stable siderophore-iron complexes. The affinity of iron to siderophores is generally higher (300-fold) than that of iron to host proteins involved in iron transport, such as transferrin and ferritin [227]. This high affinity is a key aspect of how siderophores function effectively in iron acquisition, especially in environments where iron is scarce. Bacteria utilize specific receptors to recognize and transport the siderophore-iron complex back into the cell [214]. The recognition of the complex can trigger the internalization of iron through active transport mechanisms. Upon uptake, the metal ion can be released from the siderophore inside the cell, allowing it to be utilized for various metabolic processes, including enzyme function and energy production. The intracellular presence of iron subsequently downregulates the expression of siderophore biosynthesis genes via the Fur regulatory system or other mechanisms, completing the feedback loop.

Although delivering iron into the gut may mitigate iron limitation-induced virulence in various pathogens and thus may theoretically reduce the virulence of some bacteria, it can be a double-edged sword if it also causes oxidative stress [228]. Alternative strategies focusing on enhancing the host's iron regulatory mechanisms or using antibacterial therapies may be considered without the associated risks of iron supplementation.

In summary, opposing forces of iron regulation between pathogenic bacteria's need for iron and the host's response to deprive them of this critical growth factor during a major perturbation, are an example of the challenges of maintaining molecular equipoise in the host-pathogen interaction during major physiologic stress.

4.2 Phosphate limitation during surgical stress

Phosphate limitation in the gut during surgical stress can significantly impact various physiological processes and microbial dynamics [201]. Surgical stress triggers an inflammatory response that can elevate the body's demand for phosphate. Phosphate is a crucial component of adenosine triphosphate (ATP), nucleic acids, and cell membranes. The physiologic stress associated with the process of surgery can lead to increased cellular activities that utilize phosphate, such as tissue repair and immune activation. Patients undergoing surgery may experience fluid shifts and changes in nutritional intake that can adversely affect their phosphate levels. These changes can limit the availability of dietary phosphate, especially if oral intake is restricted preoperatively or if there is inadequate nutrient absorption postoperatively [201]. Phosphate metabolism is closely related to iron metabolism in the gut [174]. Both nutrients have interconnected regulatory mechanisms that can influence each other. For instance, the presence of hepcidin, a key regulator of iron, can also affect phosphate transport and homeostasis [229]. Hormones like parathyroid hormone (PTH) and fibroblast growth factor 23 (FGF23) regulate phosphate metabolism and may be altered during surgical stress. An increase in PTH can lead to changes in kidney function that impact phosphate excretion and homeostasis [230, 231].

Phosphate is essential for many microbial processes in the gut. Phosphate limitation can affect the growth and metabolic activity of gut bacteria. Some beneficial microbes may struggle to thrive under low phosphate conditions, leading to dysbiosis, a degree of microbial disruption that itself can promote the growth of pathogenic organisms [201]. Under conditions of phosphate limitation opportunistic pathogens may gain a competitive advantage. Some pathogens possess specialized mechanisms to obtain and utilize phosphate more efficiently, allowing them to thrive in the gut even when essential nutrients are scarce.

Many bacteria, especially pathogenic strains, have evolved various mechanisms to acquire phosphate from organic and inorganic sources in their environment. As mentioned, phosphate is an essential nutrient for many cellular processes, including energy metabolism, nucleic acid synthesis, and signaling pathways. Some bacteria can utilize organic compounds such as phosphonates that contain a carbon-phosphorus (C-P) bond to acquire phosphorus [232]. This is particularly important in environments where free inorganic phosphate is limited. Enzymatic hydrolysis is a process that typically involves enzymes such as phosphonatases that cleave the C-P bond, releasing inorganic phosphate (Pi) that can then be taken up by the bacterial cell [232]. One of the most effective mechanisms is expression of extracellular enzymes known as phosphatases belonging to the group of exto-nucleotidases that hydrolyze different specific substrates including organic phosphate esters, such as nucleotides (e.g., ATP, ADP) and phospholipids, releasing inorganic phosphate for its uptake [233]. Different types of phosphatases such as alkaline phosphatases, acid phosphatases, or specific enzymes for particular organic compounds target different substrates. Upon secretion, these phosphatases act on organic phosphate substrates, converting them into free phosphate ions that can then be assimilated by the bacterial cells. In some cases, certain bacteria prefer to consume organic phosphates, where they break them down with phosphatases.

Bacteria have also evolved several regulatory mechanisms that sense phosphate availability in their environment. Systems such as the PstS- dependent expression of Pho regulon upregulate the expression of genes involved in phosphate uptake and utilization when phosphate is scarce [174, 234] and are tightly connected to the global virulence activation [174, 235]. Some clinical strains of pathogens (i.e., *P. aeruginosa*) have been shown to develop systems for extra-scavenging environmental phosphate concentrations by developing PstS-containing appendages [236, 237].

Despite the tightly regulated and common response to phosphate limitation in the microbial world, there is significant variability in the timing and concentration at which the bacterial response to phosphate is activated. For example, some strains respond earlier due to higher threshold of recognition of decreasing extracellular phosphate. Generally, phosphate concentrations in the range of 0.5 to 1 mM are considered sufficient for many bacterial species to grow and survive under normal conditions. Within this range, bacteria can carry out normal metabolic functions, including growth, replication, and biosynthesis. However, for *P. aeruginosa,* even at these physiologic levels, PstS expression is already induced. While optimal phosphate concentrations generally range from 0.5 to 1 mM, levels below 0.1 mM are often recognized as low by many bacterial species, triggering adaptive responses to enhance phosphate acquisition. Some bacteria can experience significant stress or metabolic slowdown when phosphate concentrations fall to the level of 4 mkM [238]. The ability to "sense and respond" to phosphate levels is vital for bacterial survival, especially in environments where this nutrient is limited. Variability in the threshold at which phosphate responses are triggered among bacterial species can significantly influence their ability to expand when present in the gut during stress. For example, in response to phosphate limitation, opportunistic pathogens such as *P. aeruginosa* express their global virulence machinery including to acquire organic phosphate, siderophores, quorum sensing MvfR-PQS, and efflux pumps [174]. This ability to thrive under phosphate limitation allows these pathogens to exploit available resources that commensal bacteria cannot utilize effectively; this situation inhibits growth of the commensal microflora via secreted virulence factors by pathogenic strains that include the precursors of PQS and that can directly suppress the immune response (**Figure 1**) [174].

In summary, situations in which prolonged illness and its treatment involve surgical stress or inflammatory conditions, can lead to the diminished availability of key nutrients that some bacteria use for growth and survival, while other use to express virulence with the goal of obtaining phosphate in deeper tissues. This process involves systemic phosphate regulatory mechanisms such as the release of phosphatonins, which shift the availability of phosphate away from the gut to more vital organs of survival such as the heart and brain. Opportunistic pathogens can better exploit these changing conditions due to their adaptive mechanisms and metabolic versatility. The regulation of virulence factors in response to these stressors can also affect phosphate levels such that circumstances favor opportunistic pathogens with a greater probability of survival and pathogenicity. Differences in threshold phosphate levels can benefit opportunistic pathogens by enhancing their ability to acquire iron, utilize organic phosphates, and adapt their gene expression in response to changing environments. Their capacity to outcompete commensal bacteria and form protective biofilms allows them to expand their niche colonization in the gut resulting in enhanced survival and pathogenicity, especially during stress conditions when the disease state and its treatment are intensified.

5. Basic strategies for preventing postoperative infection-related complications by targeting the intestinal microbiota

Based on the above discussion highlighting the main characteristics of pathobiomes and their behaviors during stress conditions, two main strategies could be developed to prevent and/or protect a host during stress conditions such as those that occur in postoperative infection-related complications. These include: (1) a strategy targeting the composition and function of the gut microbiota, and (2) prevention of virulence activation by opportunistic pathogens in response to conditions of host stress.

5.1 Strategy targeting composition and function of gut microbiota

Preventing postoperative infection-related complications by targeting the composition and function of gut microbiota is an innovative approach that recognizes the key role of the gut's microbiome composition and function to maintain immune homeostasis and to enhance recovery following major surgery.

5.1.1 Direct restoration of microbiota composition via fecal microbial transplant (FMT)

Fecal microbial transplantation (FMT) that involves transferring fecal matter from a healthy donor to a recipient to restore a balanced microbiome, has a long and somewhat varied history. The concept of using fecal matter to promote host health can be traced back to traditional practices in various cultures, including ancient China. Historical records suggest that the use of fecal material for therapeutic purposes dates back to at least the fourth century in China, where it was referred to as "yellow soup" (or "huang tang" in Mandarin). This mixture was used to treat severe gastrointestinal diseases, particularly diarrhea and dysentery [239]. Over the years, FMT has gained acceptance in clinical settings, with various protocols established for donor screening, preparation of fecal samples, and methods of administration (such as colonoscopy, enema, or oral capsules) [240–245]. In many countries, FMT is now considered a standard treatment for recurrent CDI, and various clinical trials are exploring its use in other conditions linked to dysbiosis, such as IBD, irritable bowel syndrome (IBS), and metabolic disorders [246–248].

In animal models involving major surgery, such as partial hepatectomy in which some mice survive while others succumb to infection, administration of an FMT demonstrated a high degree of efficacy enhancing the survival of mice in association with an improvement in the immune response, a clearing of systemic infection and recovery of the microbiota within the colon and cecal crypts [144, 249, 250].

Compared to other systemic biologic-based therapies, FMT seems to be relatively simple and can be performed using methods like colonoscopy, enemas, or oral capsules. Additionally, FMT is a cost-effective treatment compared to prolonged antibiotic therapies or other systemic treatment options. By restoring the gut flora, an FMT is intended to enhance immune responses, and enhance the competitive exclusion of pathogenic strains thus providing a potential benefit to immunocompromised patients. However, there are some challenges that make the use of FMT logistically difficult such as delivering via colonoscopy is not easy, what makes an effective donor for FMT, and gut microbiota environment that may not be favored for the FMT.

Furthermore, there are serious disadvantages to FMT such as a risk of transmitting highly pathogenic infectious agents from the donor to the recipient [251]. Although screening methods are intended to exclude donor FMTs that harbor pathogenic strains, the perceived "innocuous" donor sample may not be as innocent as originally understood. A healthy donor FMT can shift its composition and/or phenotype to an unfavorable state resulting in a fully adapted pathobiome that can cause inadvertent harm to its recipient.

As such the FMT strategy needs further development. Developing FMT strategies that focus on a prior analysis and modification of the gut environment, including metabolites, vitamins, and minerals, can enhance its effectiveness and potentially reduce its adverse effects. A comprehensive understanding of the individual patient's microbiome and nutritional needs can lead to more personalized and successful interventions.

A program in which gut microbiota is fully screened and banked from young individuals (5–6 years) whose exposures to antibiotics and other perturbing agents should be considered. Were a donor FMT to be screened, curated and biobanked and determined to be safe and efficacious, it could provide an important source for personalized support of the gut microbiota during conditions in which its recovery is impaired.

5.1.2 Resource strategy: Improving microbiota composition via dietary modification

Recent experiments in animal models have demonstrated that a preoperative course of dietary preparation (DietPreHab) can lead to improved postoperative conditions and perhaps limit postoperative infection-related complications [252–254]. The widespread intake of a so-called Western diet (WD), high in fat, high in sugar, low in fiber, may be a major unrecognized risk factor for postoperative complications [255–257]. Its chronic exposure, common in Westernized societies, can shift the microbiome toward a pathobiome characterized, in part, by a sharp decrease in butyrate levels [115, 258]. For example, following 6 weeks on a Western diet, mice become more susceptible to surgical stress in the form of worse survival compared to their littermate chow fed mice. However, a short course of preoperative switching to a healthy high-fiber standard rodent chow-based diet restores their stability and increases survival after a partial liver resection, even when accompanied by antibiotic treatment [257].

The concept of DietPreHab refers to nutritional optimization and dietary preparation of patients before undergoing surgery beyond the usual systemic markers of visceral protein status such as albumin and total protein [259, 260]. This approach aims to improve postoperative outcomes by ensuring that patients are in the best possible nutritional status inclusive of markers that enhance the nutritional profile of their microbiome. Increasing fiber, protein, vitamin and minerals intake prior to surgery improves the composition and function of gut microbiota which plays a vital role in metabolic health and immune responses, and reduce systemic inflammation, leading to less postoperative infection-related complications. Enhancing the composition and function (i.e., immune-stimulating metabolites) of the microbiota such that proteins, vitamins, and minerals are optimized is an essential target of efficacy to ensure microbiota "readiness" for surgery [260]. In addition, in order to facilitate healing of any operative wound, it is essential that collagen breakdown and synthesis remains balanced. Therefore, minimizing the appearance of collagenolytic pathogens such as *E. faecalis*, *S. marcescens*, *P. mirabilis* is important as their mere presence can lead to impaired wound healing post-surgery [93].

This protective approach can significantly contribute to a patient's overall surgical outcome, highlighting the importance of holistic-based nutrition in the perioperative setting to not only target host cellular systems but also its microbiota. However, the effective communication and individualized approaches are essential for the successful implementation of dietary prehabilitation programs.

An important component of pre- operative program is a pre-operative dental care that needs to be performed to prevent oral pathogen-based postoperative infection and inflammation [261].

5.1.3 Resource strategy: Providing metabolites of healthy microbiota or their precursors

Providing metabolites from the healthy microbiota or their precursors is an emerging strategy to improve postoperative outcomes. Key metabolites and precursors that can be beneficial in the context of postsurgical recovery are SCFAs, especially butyrate [262–264], and indoles including its dietary precursor tryptophan [41, 265]. Importantly both are exclusively produced by the gut microbiota.

Butyrate is produced primarily through the fermentation of dietary fibers by beneficial gut bacteria. It is a primary energy source for colonic epithelial cells and plays a vital role in maintaining gut barrier integrity [266]. It also reduces inflammation by promoting the production of anti-inflammatory cytokines and regulating immune cell function [267]. Butyrate is known as a potent inhibitor of histone deacetylases (HDACs) [268, 269]. This inhibition leads to an increase in histone acetylation, which can alter gene expression patterns by making DNA more accessible for transcription. The modulation of HDAC activity can influence the expression of various genes involved in immune responses, including cytokines, chemokines, and surface receptors, enhancing anti-inflammatory pathways and promoting regulatory T-cell (Treg) development [270].

Butyrate strengthens the gut barrier by enhancing the function and proliferation of colonic epithelial cells [271]. By maintaining tight junction integrity, butyrate helps prevent alterations in the permselectivity of the gut which can lead to systemic infection [272].

However, direct administration of butyrate to the gut is not a common practice and faces many challenges [273]. Not only is it metabolized by neighboring bacteria, gut epithelial cells and the liver, but depending on how it is administered, can lead to localized irritation, inflammation, or even ulceration if the local concentration administered is excessive. Alternative routes of butyrate administration such as intrarectal delivery are often complicated for patients with chronic disorders. Butyrate supplements, such as sodium butyrate or calcium magnesium butyrate, are available in forms that are designed to release butyrate slowly in the gut, reducing the risk of irritation. Yet, despite its many benefits, while it seems axiomatic that if a nutrient is low and is associated with a poor outcome, its supplementation must somehow be considered beneficial, with butyrate, this type of syllogistic reasoning is not clear.

The delivery and consumption of dietary tryptophan may be another important consideration to rehabilitate the gut microbiota prior to surgery [42]. In a mouse model, oral tryptophan delivery was shown as effective strategy to prevent mortality from a systemic infection [42]. Adding tryptophan to the drinking water a week before a systemic infection in mice enhanced survival via mechanisms mediated by an increase in the production of indoles. The microbial metabolites of tryptophan, (i.e., indole-3-acetic acid, tryptophol, indole-3-propionic acid, etc.) can bind to the AhR receptor and trigger a phenotypic shift in macrophages to become more M2-like and exhibit anti-inflammatory activity with increased phagocytic activity leading to the clearance of systemic infections. Further, clinical analyses of stool and serum of patients demonstrated a correlation of decreased concentrations of indoles with a negative outcome [42].

If one were to consider the enrichment of tryptophan concentration in gut with the goal of enhancing the "readiness" of the microbiota for surgery, then dietary sources such as food enriched with tryptophan (turkey, chicken fish, cheese, nuts, seeds, eggs, soybeans) might be demonstrated to be beneficial. Tryptophan supplements can help individuals who may not get enough tryptophan from their diet. Targeted delivery such as a microencapsulation techniques can be used to effectively deliver tryptophan through the gastrointestinal tract [274]. However, it is important to note that only certain bacteria convert tryptophan into indoles, therefore, the composition of the gut microbiota itself can determine the efficacy of tryptophan supplementation. Providing tryptophan-metabolizing bacteria to the gut in high-risk patients requiring major surgery (*Bacteroides*, *Lactobacillus*, and *Bifidobacterium*) may be a complementary option to tryptophan delivery. We noted, similar to butyrate that oral delivery of indoles is not effective in animal models of systemic infection and again may be due to its degradation in the gastrointestinal tract by neighboring bacteria [42] and many other factors. Although medicinal pharmacology is often based on identifying a deficiency and restoring that deficiency with supplementation of various forms of the identified agent, this process of reasoning may not be applicable to those factors identified in the microbiome.

Finally, it is important to note the potential synergism between tryptophan and butyrate [119]. Tryptophan and butyrate can collectively influence the composition of the gut microbiota. A diet rich in tryptophan may promote the growth of specific bacteria that are also capable of producing butyrate, leading to enhanced SCFA levels in the gut [119]. The presence of butyrate and other SCFAs can modulate the metabolism of tryptophan in the gut, promoting beneficial microbial interactions and enhancing the production of metabolites linked to gut health. For example, by the inhibition of indoleamine 2,3-dioxygenase activity, butyrate suppresses kynurenine production from tryptophan [275]. Furthermore, carbohydrates such as resistant starch, facilitates the carbohydrate metabolism, leading to increased SCFA production while reducing the rate of tryptophan degradation and indoles [276]. Butyrate is a source of energy for colonocytes (the epithelial cells of the colon) and plays a vital role in maintaining gut barrier integrity. Tryptophan supports cellular health and may work synergistically with butyrate to enhance epithelial function. Both tryptophan and butyrate contribute to maintaining tight junctions between epithelial cells — a critical process by which intestinal permeability is maintained [266, 277]. Taken together, these findings support the consumption of fiber-rich and tryptophan-rich diets as methods to modulate the gut microbiome in the pre-surgical period.

In summary, the above discussion to develop strategies targeting both the composition and function of the gut microbiota, such as with an FMT and/or pre-operative dietary modifications, is gaining attention in improving health outcomes. This may be particularly useful in the context of surgery and recovery. Strategies such as fecal microbiota transplantation (FMT) and pre-operative diet modifications are promising approaches that have not yet been tested for their efficacy in the standard care of a patient anticipating a major surgical intervention. Such an approach may have many benefits over the expanded use of antibiotics that justify their overuse with the reasoning that "if some is good, more must be better."

6. Strategies focused on the suppression of negative effect of gut pathobiome during surgical stress

It is important to note that gut microbiome-related strategies focused on corrections of their composition and function, may require an extended period to adequately modify and stabilize a healthy gut microbiome prior to surgery. Given that most surgery is elective and is often delayed due to intercurrent therapies such as neoadjuvant chemotherapy or cardiac rehabilitation, sufficient time may be allowed to render the gut microbiota "ready" for surgery. In the case of emergency surgery, how the gut microbiota is treated as the patient recovers may need redirection. Two main tasks should be considered: (1) protection of erosion of gut epithelial barrier, a function dependent on an intact and healthy microbiome and (2) prevention of virulence activation by pathogenic species that are selected for as a result of antibiotic use and the feeding of chemically define, fiber free, sterile diets.

6.1 Protection of epithelial barrier during surgery

The effects of surgery on the gut epithelial barrier are profound, leading to increased permeability, inflammation, and changes in gut microbiota that complicate postoperative recovery. Increased permeability of the gut barrier is purported to allow the potential translocation of pathogens and endotoxins into the bloodstream, leading to systemic infections, which are a risk factor for an endogenous bacteremia or toxemia what many have termed "gut-derived sepsis" [278–281]. Many have argued that if the tight junctions of the lining gut epithelium could be rendered "tighter" such that bacteria and toxins from the gut do not activate inflammation, the outcome can be improved [282].

An innovative approach has been suggested to prevent this occurrence by creating an artificial mucus-like layer that could cover gut intestinal epithelium preventing adsorption of intestinal pathogens and their toxins [283–285]. Among different types of mucus-like layers, one represents a specific triblock structure of high molecular weight polyethylene glycol consisting of short hydrophobic region linked to long hydrophilic chains PEG15-20 (Sigma) [286]. PEG15-20 creates a "repulsive" effect between the epithelium and microbial pathogens [286]. Mechanism of repulsive effect was revealed by the synchrotron small-angle X-ray scattering analysis that demonstrated that PEG15-20 binds to the lipid bilayer in such a way that it directs two symmetrical hydrophilic blocks of PEG15-20 laterally across the membrane-water interface, forming a polymer coating on the lipid bilayer (**Figure 2**) [287].

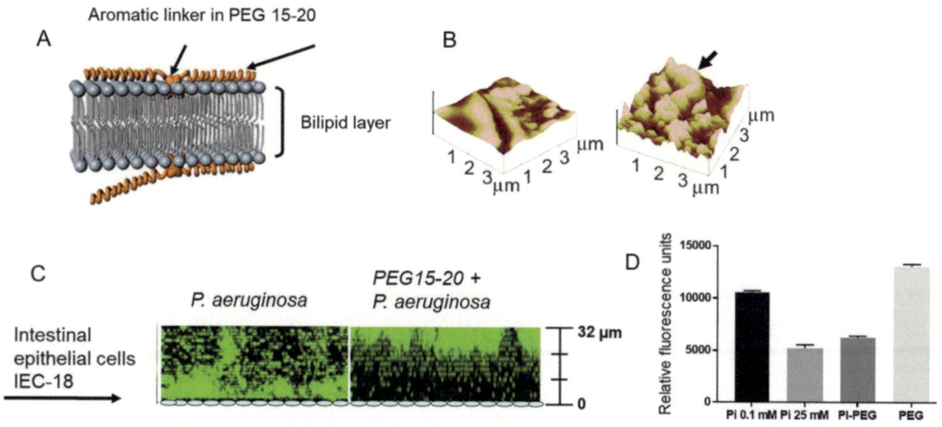

Figure 2.
ABA-PEG20k-Pi20 possesses a dual protective effect of ABA-PEG and bound phosphate groups. This figure is created based on the findings described in Refs. Valuckaite et al. [287], and Mao et al. [288]. (A). Model based on Synchrotron small-angle X-ray scattering analysis that demonstrated that PEG15-20 binds to the lipid bilayer in such a way that it directs two symmetrical hydrophilic blocks of PEG15-20 laterally across the membrane-water interface, forming a polymer coating on the lipid bilayer. (B). Atomic force microscopy (AFM) of the surface of intestinal segments in mice where PEG15-20 is included in drinking water demonstrated that PEG15-20 increases the protective layer on the intestinal muco-epithelial surface. (C). The repellant effect of P. aeruginosa *from monolayers of intestinal epithelial cells.* P. aeruginosa *cells are no longer unable to attach onto the intestinal cells.* P. aeruginosa *constitutively expressing green fluorescent protein EGFP was used in these experiments. (D). Pi-PEG (ABA-PEG20k-Pi20) but not a parent non-phosphorylated compound (PEG) is able to suppress the activation of PstS similar to that by inorganic phosphate at 25 mM. Panels A-C are reprinted in part with permission from Valuckaite et al. [287]. Copyright 2019. American Journal of Physiolology Gastrointestinal Liver Physioliology. Panel D is reprinted in part with permission from Mao et al. [288]. Copyright 2017. American Chemical Society.*

The hydrophobic block can anchor below the lipid headgroup region and thus can anchor the polymer to the membrane (**Figure 2**) [287]. This creates an ideal conformational state for PEG15-20 to form a barrier and protect the epithelium.

Atomic force microscopy (AFM) of the surface of intestinal segments in mice where PEG15-20 was included in drinking water demonstrated that PEG15-20 may form an additional protective layer on the intestinal muco-epithelial surface (**Figure 2**) [287]. The AFM analysis of bacterial cells also revealed a layer of PEG15-20 surrounding bacterial cells [286]. This may explain its repellant effect as bacterial cells are unable to attach onto the intestinal epithelium and transduce signals leading to altered permselectivity.

Furthermore, PEG15-20 and similar molecules may preserve the integrity of the intestinal epithelium following exposure to barrier disrupting factors such as the effect of secondary bile acids or irradiation [287, 289, 290]. The inclusion of PEG15-20 in the preservative solution for organ transplantation protects the integrity of the epithelial layer and preserves the vascular system of organs [291].

To find out if orally delivered PEG15-20 could reach distal colon, a fluorescein-labeled PEG15-20 compound was created. Findings in a rat model demonstrated that this compound was able to remain durably embedded to cover the intestinal epithelium all the way down to colon, and was gradually extracted with feces without penetrating into the systemic compartment [287]. In addition, rodents drinking PEG solution were protected against lethal gut-derived sepsis in a model in which P. aeruginosa was introduced in in mice subjected to partial hepatectomy [286].

6.2 Phosphorylated triblock high-molecular-weight polymer that exploits the known properties of phosphate (Pi) and polyethylene glycol to suppress microbial virulence and protect the integrity of the intestinal epithelium

As discussed above, limitations in extracellular iron and phosphate are critical nutrients in the host-microbiota-pathogen interface. Providing iron and phosphate as direct oral solutions during surgical stress while seemingly beneficial, are challenged by issues of absorption, tolerance, and disruptions in metabolic regulation. Therefore, PEG15-20 was chosen as a carrier of phosphate bound to the PEG backbone via its hydroxyl groups on its hydrophilic chains (**Figure 2**).

The *de novo* synthesized compound ABA-PEG20k-Pi20 possesses a dual protective effect in the host-microbiome-pathogen interaction whereby the PEG is cytoprotective to the intestinal epithelium and its covalent bonding to phosphate provides bacteria with a much needed nutrient that can suppress multiple virulence pathways that are triggered by phosphate limitation, iron limitation and host stress signals that induce conserved quorum sensing circuits across a variety of pathogens relevant to several host species [288]. Furthermore, delivering phosphate on the phosphorylated molecule similar to the delivery on inorganic phosphate suppresses the activation of both siderophore induction even at the low concentration of iron and quorum sensing activation induced by host stress signals such as opioids [174, 288, 292, 293].

Pi-PEG has demonstrated efficacy in multiple animal models of postoperative infection-related complications. For example, ABA-PEG20k-Pi20 suppresses collagenase production in several collagenolytic pathogens, including *E. faecalis*, *S. marcescens* and *P. aeruginosa,* and prevents colonic anastomotic leak in mice [294]. By suppressing the collagenolytic activity of pathogens, ABA-PEG20k-Pi20 also prevents colon cancer recurrence in mice [256]. Importantly, suppression of collagenase production may be a more evolutionarily stable approach compared to expanded application of antibiotics, which carry the risk of antibiotic resistance. For example, antibiotic treatment focused on eliminating one strain of bacteria selected for the colonization of anastomotic tissues by another collagenolytic pathogen, while an expanded multi-drug antibiotic combination directed against both gram-positive and gram-negative pathogens led to the colonization of anastomotic tissue by collagenolytic fungi [256]. As such, replacement by, or selection for, renegade pathogens when treating tissues with antibiotics may not be efficacious in the long run. Oral ABA-PEG20k-Pi20 can prevent surgical site infections [295]. However, before this approach can be considered clinically efficacious and safe, clinical trials are needed.

7. Conclusion

In conclusion, strategies aimed at suppressing the negative effects of an emerging gut pathobiome over the course of surgical injury and during its recovery phase may need be considered in addition to the current utilization of antibiotics. FMT, dietary modifications, anti-inflammatory management, and metabolite supplementation may be future strategies to pursue before, during, and after a major surgical injury. Given that patients often present for elective surgery with major medical co-morbidities, an opportunity to pre-emptively intervene, such as imposing a period of dietary pre-habilitation, should be considered. Protection of epithelial barrier during

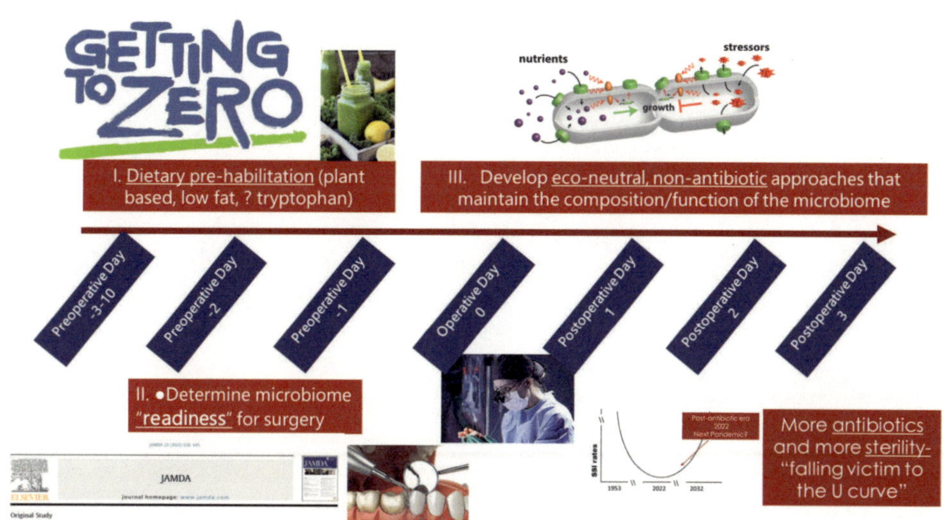

Figure 3.
Pre- and post-operative therapy to improve post-operative output. Current strategies to reduce postoperative infectious complications have successfully reduced their occurrence, but newer strategies need to be undertaken to completely eliminate all infectious complication to zero. I. Dietary pre-habilitation begins at 3-10 days prior to surgical operation. It includes plant-based diet enriched by tryptophan source and containing low fat. Additionally, pre-operative dental care is needed to prevent oral pathogens-based post-operative infection and inflammation. II. Microbiome "readiness" is determined by both the presence of health-promoting species and their production of health relevant metabolites (i.e., SCFAs, bile acids, and indole metabolites). III. Treatment strategies that alter gut nutrients (i.e., PiPEG) create a gut environment that maintains the mucus barrier, protects intestinal epithelial cells, reduces pathogen virulence and their response to host stress, and promotes homeostasis of essential bacterial communities. By utilizing novel, microbiota directed strategies we may avoid the current antimicrobial approaches that are not only having diminishing returns, but resulting in antimicrobial resistance that have the potential to undo the impact antibiotics have had on reducing infectious complications.

surgery and suppression of the virulence signaling platform during host stress with compounds like ABA-PEG20k-Pi20 is yet another/an additional strategy to pursue that silence opportunistic pathogens within pathobiomes and blocks their penetration into systemic compartment (**Figure 3**).

Acknowledgements

This work was supported by NIH grant R01GMO62344-22 (John Alverdy).

Disclosure

OZ, AZ, and JA work for Covira Surgical, Inc., a company created for the development of Pi-PEG as a drug for prevention of post-surgical infections.

Author details

Olga Zaborina*, Alexander Zaborin, Robert Keskey and John Alverdy*
Department of Surgery, University of Chicago, Chicago, IL, USA

*Address all correspondence to: olgazaborina2@gmail.com and jalverdy@surgery.bsd.uchicago.edu

References

[1] Fujisaka S, Watanabe Y, Tobe K. The gut microbiome: A core regulator of metabolism. The Journal of Endocrinology. 2023;**256**(3)

[2] Rinninella E et al. What is the healthy gut microbiota composition? A changing ecosystem across age, environment, diet, and diseases. Microorganisms. 2019;**7**(1)

[3] Wexler AG, Goodman AL. An insider's perspective: Bacteroides as a window into the microbiome. Nature Microbiology. 2017;**2**:17026

[4] James KR et al. Distinct microbial and immune niches of the human colon. Nature Immunology. 2020;**21**(3):343-353

[5] Jensen BAH et al. Small intestine vs. colon ecology and physiology: Why it matters in probiotic administration. Cell Reports Medicine. 2023;**4**(9):101190

[6] Lozupone CA et al. Diversity, stability and resilience of the human gut microbiota. Nature. 2012;**489**(7415):220-230

[7] Fassarella M et al. Gut microbiome stability and resilience: Elucidating the response to perturbations in order to modulate gut health. Gut. 2021;**70**(3):595-605

[8] Makki K et al. The impact of dietary fiber on gut microbiota in host health and disease. Cell Host & Microbe. 2018;**23**(6):705-715

[9] Fu J et al. Dietary fiber intake and gut microbiota in human health. Microorganisms. 2022;**10**(12)

[10] Wynne AG et al. An in vitro assessment of the effects of broad-spectrum antibiotics on the human gut microflora and concomitant isolation of a *Lactobacillus plantarum* with anti-Candida activities. Anaerobe. 2004;**10**(3):165-169

[11] Lekang K et al. Effects of different amoxicillin treatment durations on microbiome diversity and composition in the gut. PLoS One. 2022;**17**(10):e0275737

[12] Buffie CG et al. Profound alterations of intestinal microbiota following a single dose of clindamycin results in sustained susceptibility to *Clostridium difficile*-induced colitis. Infection and Immunity. 2012;**80**(1):62-73

[13] Elvers KT et al. Antibiotic-induced changes in the human gut microbiota for the most commonly prescribed antibiotics in primary care in the UK: A systematic review. BMJ Open. 2020;**10**(9):e035677

[14] Ng KM et al. Recovery of the gut microbiota after antibiotics depends on host diet, community context, and environmental reservoirs. Cell Host & Microbe. 2019;**26**(5):650-665 e4

[15] Lloyd-Price J, Abu-Ali G, Huttenhower C. The healthy human microbiome. Genome Medicine. 2016;**8**(1):51

[16] Suez J et al. Post-antibiotic gut mucosal microbiome reconstitution is impaired by probiotics and improved by autologous FMT. Cell. 2018;**174**(6):1406-1423 e16

[17] Schmid DW et al. A framework for testing the impact of co-infections on host gut microbiomes. Animal Microbiome. 2022;**4**(1):48

[18] Zhang F et al. Gut microbiota in COVID-19: Key microbial changes,

potential mechanisms and clinical applications. Nature Reviews. Gastroenterology & Hepatology. 2023;**20**(5):323-337

[19] Martin R et al. *Faecalibacterium*: A bacterial genus with promising human health applications. FEMS Microbiology Reviews. 2023;**47**(4)

[20] Hemarajata P, Versalovic J. Effects of probiotics on gut microbiota: Mechanisms of intestinal immunomodulation and neuromodulation. Therapeutic Advances in Gastroenterology. 2013;**6**(1):39-51

[21] Chandrasekaran P, Weiskirchen S, Weiskirchen R. Effects of probiotics on gut microbiota: An overview. International Journal of Molecular Sciences. 2024;**25**(11)

[22] Huang F et al. Postoperative probiotics administration attenuates gastrointestinal complications and gut microbiota dysbiosis caused by chemotherapy in colorectal cancer patients. Nutrients. 2023;**15**(2)

[23] Szajewska H et al. Antibiotic-perturbed microbiota and the role of probiotics. Nature Reviews. Gastroenterology & Hepatology. 2025;**22**(3):155-172

[24] Sonnenburg ED, Sonnenburg JL. Starving our microbial self: The deleterious consequences of a diet deficient in microbiota-accessible carbohydrates. Cell Metabolism. 2014;**20**(5):779-786

[25] Morrison DJ, Preston T. Formation of short chain fatty acids by the gut microbiota and their impact on human metabolism. Gut Microbes. 2016;**7**(3):189-200

[26] Fusco W et al. Short-chain fatty-acid-producing bacteria: Key components of the human gut microbiota. Nutrients. 2023;**15**(9)

[27] Shin Y et al. Roles of short-chain fatty acids in inflammatory bowel disease. Nutrients. 2023;**15**(20)

[28] Du Y et al. The role of short chain fatty acids in inflammation and body health. International Journal of Molecular Sciences. 2024;**25**(13)

[29] Silva YP, Bernardi A, Frozza RL. The role of short-chain fatty acids from gut microbiota in gut-brain communication. Frontiers in Endocrinology (Lausanne). 2020;**11**:25

[30] Hill MJ. Intestinal flora and endogenous vitamin synthesis. European Journal of Cancer Prevention. 1997;**6**(Suppl. 1):S43-S45

[31] Guetterman HM et al. Vitamin B-12 and the gastrointestinal microbiome: A systematic review. Advances in Nutrition. 2022;**13**(2):530-558

[32] LeBlanc JG et al. Bacteria as vitamin suppliers to their host: A gut microbiota perspective. Current Opinion in Biotechnology. 2013;**24**(2):160-168

[33] Levit R et al. Folate-producing lactic acid bacteria reduce inflammation in mice with induced intestinal mucositis. Journal of Applied Microbiology. 2018;**125**(5):1494-1501

[34] Levit R et al. Beneficial effect of a mixture of vitamin-producing and immune-modulating lactic acid bacteria as adjuvant for therapy in a recurrent mouse colitis model. Applied Microbiology and Biotechnology. 2019;**103**(21-22):8937-8945

[35] Krautkramer KA, Fan J, Backhed F. Gut microbial metabolites as multi-kingdom intermediates. Nature Reviews. Microbiology. 2021;**19**(2):77-94

[36] Soto-Martin EC et al. Vitamin biosynthesis by human gut butyrate-producing bacteria and cross-feeding in synthetic microbial communities. MBio. 2020;**11**(4)

[37] Neis EP, Dejong CH, Rensen SS. The role of microbial amino acid metabolism in host metabolism. Nutrients. 2015;7(4):2930-2946

[38] Li TT et al. Microbiota metabolism of intestinal amino acids impacts host nutrient homeostasis and physiology. Cell Host & Microbe. 2024;**32**(5):661-675 e10

[39] Lopez MJ, Mohiuddin SS. Biochemistry, essential amino acids. In: StatPearls. Treasure Island (FL); 2025

[40] Dalangin R, Kim A, Campbell RE. The role of amino acids in neurotransmission and fluorescent tools for their detection. International Journal of Molecular Sciences. 2020;**21**(17)

[41] Li X et al. New insights into gut-bacteria-derived indole and its derivatives in intestinal and liver diseases. Frontiers in Pharmacology. 2021;**12**:769501

[42] Keskey RC et al. Enterobactin inhibits microbiota-dependent activation of AhR to promote bacterial sepsis in mice. Nature Microbiology. 2025;**10**(2):388-404

[43] Konopelski P, Ufnal M. Indoles - Gut bacteria metabolites of tryptophan with pharmacotherapeutic potential. Current Drug Metabolism. 2018;**19**(10):883-890

[44] Kaur H, Bose C, Mande SS. Tryptophan metabolism by gut microbiome and gut-brain-axis: An in silico analysis. Frontiers in Neuroscience. 2019;**13**:1365

[45] Hubbard TD, Murray IA, Perdew GH. Indole and tryptophan metabolism: Endogenous and dietary routes to AH receptor activation. Drug Metabolism and Disposition. 2015;**43**(10):1522-1535

[46] Elias AE et al. Activation of the aryl hydrocarbon receptor via indole derivatives is a common feature in skin bacterial isolates. Journal of Applied Microbiology. 2024;**135**(11)

[47] Kim S, Seo SU, Kweon MN. Gut microbiota-derived metabolites tune host homeostasis fate. Seminars in Immunopathology. 2024;**46**(1-2):2

[48] Gutierrez-Vazquez C, Quintana FJ. Regulation of the immune response by the aryl hydrocarbon receptor. Immunity. 2018;**48**(1):19-33

[49] Wang G et al. Microbiota-derived indoles alleviate intestinal inflammation and modulate microbiome by microbial cross-feeding. Microbiome. 2024;**12**(1):59

[50] Ye X et al. Dual role of indoles derived from intestinal microbiota on human health. Frontiers in Immunology. 2022;**13**:903526

[51] Hassan SM et al. Indole compounds in oncology: Therapeutic potential and mechanistic insights. Pharmaceuticals (Basel). 2024;**17**(7)

[52] Dietrich C. Antioxidant functions of the aryl hydrocarbon receptor. Stem Cells International. 2016;**2016**:7943495

[53] Sittipo P et al. The function of gut microbiota in immune-related neurological disorders: A review. Journal of Neuroinflammation. 2022;**19**(1):154

[54] Zhou Y et al. The role of the indoles in microbiota-gut-brain axis

and potential therapeutic targets: A focus on human neurological and neuropsychiatric diseases. Neuropharmacology. 2023;**239**:109690

[55] Lee JH, Lee J. Indole as an intercellular signal in microbial communities. FEMS Microbiology Reviews. 2010;**34**(4):426-444

[56] Lee JH, Wood TK, Lee J. Roles of indole as an interspecies and interkingdom signaling molecule. Trends in Microbiology. 2015;**23**(11):707-718

[57] Krezalek MA et al. The shift of an intestinal "microbiome" to a "Pathobiome" governs the course and outcome of sepsis following surgical injury. Shock. 2016;**45**(5):475-482

[58] Munley JA et al. Multicompartmental traumatic injury and the microbiome: Shift to a pathobiome. Journal of Trauma and Acute Care Surgery. 2023;**94**(1):15-22

[59] Mohammed LI et al. Smoking induced salivary microbiome dysbiosis and is correlated with lipid biomarkers. BMC Oral Health. 2024;**24**(1):608

[60] Chakladar J et al. The intratumor bacterial and fungal microbiome is characterized by HPV, smoking, and alcohol consumption in head and neck squamous cell carcinoma. International Journal of Molecular Sciences. 2022;**23**(21)

[61] Newsome R, Yang Y, Jobin C. Western diet influences on microbiome and carcinogenesis. Seminars in Immunology. 2023;**67**:101756

[62] Zinocker MK, Lindseth IA. The Western diet-microbiome-host interaction and its role in metabolic disease. Nutrients. 2018;**10**(3)

[63] Lee I et al. Proton pump inhibitors increase the risk of carbapenem-resistant *Enterobacteriaceae* colonization by facilitating the transfer of antibiotic resistance genes among bacteria in the gut microbiome. Gut Microbes. 2024;**16**(1):2341635

[64] Imhann F et al. Proton pump inhibitors affect the gut microbiome. Gut. 2016;**65**(5):740-748

[65] Huang EY et al. Using corticosteroids to reshape the gut microbiome: Implications for inflammatory bowel diseases. Inflammatory Bowel Diseases. 2015;**21**(5):963-972

[66] Morrison PK et al. The equine gastrointestinal microbiome: Impacts of age and obesity. Frontiers in Microbiology. 2018;**9**:3017

[67] Kamel M et al. Microbiome dynamics: A paradigm shift in combatting infectious diseases. Journal of Personalized Medicine. 2024;**14**(2)

[68] Alverdy JC, Krezalek MA. Collapse of the microbiome, emergence of the Pathobiome, and the immunopathology of sepsis. Critical Care Medicine. 2017;**45**(2):337-347

[69] Shogan BD et al. Intestinal anastomotic injury alters spatially defined microbiome composition and function. Microbiome. 2014;**2**:35

[70] Defazio J et al. The opposing forces of the intestinal microbiome and the emerging pathobiome. The Surgical Clinics of North America. 2014;**94**(6):1151-1161

[71] Opstelten JL et al. Gut microbial diversity is reduced in smokers with Crohn's disease. Inflammatory Bowel Diseases. 2016;**22**(9):2070-2077

[72] Manichanh C et al. Reduced diversity of faecal microbiota in Crohn's disease revealed by a metagenomic approach. Gut. 2006;**55**(2):205-211

[73] Khan I et al. Alteration of gut microbiota in inflammatory bowel disease (IBD): Cause or consequence? IBD treatment targeting the gut microbiome. Pathogens. 2019;**8**(3)

[74] Tsai YC et al. Alternations of the gut microbiota and the Firmicutes/Bacteroidetes ratio after biologic treatment in inflammatory bowel disease. Journal of Microbiology, Immunology, and Infection. 2025;**58**(1):62-69

[75] Shim JO. Gut microbiota in inflammatory bowel disease. Pediatric Gastroenterology, Hepatology & Nutrition. 2013;**16**(1):17-21

[76] Magne F et al. The firmicutes/bacteroidetes ratio: A relevant marker of gut dysbiosis in obese patients? Nutrients. 2020;**12**(5)

[77] Turnbaugh PJ et al. An obesity-associated gut microbiome with increased capacity for energy harvest. Nature. 2006;**444**(7122):1027-1031

[78] Hall KD et al. The energy balance model of obesity: Beyond calories in, calories out. The American Journal of Clinical Nutrition. 2022;**115**(5):1243-1254

[79] Barlow GM, Mathur R. Type 2 diabetes and the microbiome. Journal of the Endocrine Society. 2022;**7**(2):bvac184

[80] Scheithauer TPM et al. Gut microbiota as a trigger for metabolic inflammation in obesity and type 2 diabetes. Frontiers in Immunology. 2020;**11**:571731

[81] Liu Y, Dai M. Trimethylamine N-oxide generated by the gut microbiota is associated with vascular inflammation: New insights into atherosclerosis. Mediators of Inflammation. 2020;**2020**:4634172

[82] Xiong RG et al. The role of gut microbiota in anxiety, depression, and other mental disorders as well as the protective effects of dietary components. Nutrients. 2023;**15**(14)

[83] Clapp M et al. Gut microbiota's effect on mental health: The gut-brain axis. Clinics and Practice. 2017;**7**(4):987

[84] Kelly MS et al. The environmental microbiome, allergic disease, and asthma. The Journal of Allergy and Clinical Immunology. In Practice. 2022;**10**(9):2206-2217 e1

[85] Valverde-Molina J, Garcia-Marcos L. Microbiome and asthma: Microbial dysbiosis and the origins, phenotypes, persistence, and severity of asthma. Nutrients. 2023;**15**(3)

[86] Zaborin A et al. Membership and behavior of ultra-low-diversity pathogen communities present in the gut of humans during prolonged critical illness. MBio. 2014;**5**(5):e01361-e01314

[87] Agudelo-Ochoa GM et al. Gut microbiota profiles in critically ill patients, potential biomarkers and risk variables for sepsis. Gut Microbes. 2020;**12**(1):1707610

[88] Salameh TJ et al. Gut microbiome dynamics and associations with mortality in critically ill patients. Gut Pathogens. 2023;**15**(1):66

[89] Otani S, Chihade DB, Coopersmith CM. Critical illness and the role of the microbiome. Acute Medicine & Surgery. 2019;**6**(2):91-94

[90] Bass D et al. The pathobiome in animal and plant diseases.

Trends in Ecology & Evolution. 2019;**34**(11):996-1008

[91] Baumler AJ, Sperandio V. Interactions between the microbiota and pathogenic bacteria in the gut. Nature. 2016;**535**(7610):85-93

[92] Chow J, Tang H, Mazmanian SK. Pathobionts of the gastrointestinal microbiota and inflammatory disease. Current Opinion in Immunology. 2011;**23**(4):473-480

[93] Hyoju SK et al. Low-fat/high-fibre diet prehabilitation improves anastomotic healing via the microbiome: An experimental model. The British Journal of Surgery. 2020;**107**(6):743-755

[94] Guyton K, Alverdy JC. The gut microbiota and gastrointestinal surgery. Nature Reviews. Gastroenterology & Hepatology. 2017;**14**(1):43-54

[95] Warraich F et al. Factors associated with fecal microbiota transplant failure in the treatment of recurrent *Clostridioides difficile* infection: A single-center retrospective study. Cureus. 2023;**15**(9):e45118

[96] Yadegar A et al. Fecal microbiota transplantation: Current challenges and future landscapes. Clinical Microbiology Reviews. 2024;**37**(2):e0006022

[97] Seekatz AM, Young VB. *Clostridium difficile* and the microbiota. The Journal of Clinical Investigation. 2014;**124**(10):4182-4189

[98] Song JH, Kim YS. Recurrent *Clostridium difficile* infection: Risk factors, treatment, and prevention. Gut Liver. 2019;**13**(1):16-24

[99] Gilliland A et al. Pathobionts in inflammatory bowel disease: Origins, underlying mechanisms, and implications for clinical care. Gastroenterology. 2024;**166**(1):44-58

[100] Santos JM et al. Pre and post-operative alterations of the gastrointestinal microbiome following bariatric surgery. Cureus. 2021;**13**(2):e13057

[101] Ciobarca D et al. Bariatric surgery in obesity: Effects on gut microbiota and micronutrient status. Nutrients. 2020;**12**(1)

[102] Petersen ME et al. Combination therapy delays antimicrobial resistance after adaptive laboratory evolution of *Staphylococcus aureus*. Antimicrobial Agents and Chemotherapy. 2025:e0148324

[103] Chernevskaya E et al. Serum and fecal profiles of aromatic microbial metabolites reflect gut microbiota disruption in critically ill patients: A prospective observational pilot study. Critical Care. 2020;**24**(1):312

[104] Facchin S et al. Short-chain fatty acids and human health: From metabolic pathways to current therapeutic implications. Life (Basel). 2024;**14**(5)

[105] Sereti I et al. Impaired gut microbiota-mediated short-chain fatty acid production precedes morbidity and mortality in people with HIV. Cell Reports. 2023;**42**(11):113336

[106] Liu F et al. Orthopedic surgery causes gut microbiome dysbiosis and intestinal barrier dysfunction in prodromal Alzheimer disease patients: A prospective observational cohort study. Annals of Surgery. 2022;**276**(2):270-280

[107] Komorniak N et al. Association between fecal microbiota, SCFA, gut integrity markers and depressive symptoms in patients treated in the past

with bariatric surgery-the cross-sectional study. Nutrients. 2022;**14**(24)

[108] Xu X et al. Gut microbial metabolite short-chain fatt acids partially reverse surgery and anesthesia-induced behavior deficits in C57BL/6J mice. Frontiers in Neuroscience. 2021;**15**:664641

[109] Yamada T et al. Rapid and sustained long-term decrease of fecal short-chain fatty acids in critically ill patients with systemic inflammatory response syndrome. JPEN Journal of Parenteral and Enteral Nutrition. 2015;**39**(5):569-577

[110] Valdes-Duque BE et al. Stool short-chain fatty acids in critically ill patients with sepsis. Journal of the American College of Nutrition. 2020;**39**(8):706-712

[111] Guinan J et al. Antibiotic-induced decreases in the levels of microbial-derived short-chain fatty acids correlate with increased gastrointestinal colonization of *Candida albicans*. Scientific Reports. 2019;**9**(1):8872

[112] Maltz RM et al. Social stress affects colonic inflammation, the gut microbiome, and short-chain fatty acid levels and receptors. Journal of Pediatric Gastroenterology and Nutrition. 2019;**68**(4):533-540

[113] Han W et al. Disordered gut microbiota and changes in short-chain fatty acids and inflammatory processes in stress-vulnerable mice. Journal of Neuroimmunology. 2023;**383**:578172

[114] Ney LM et al. Short chain fatty acids: Key regulators of the local and systemic immune response in inflammatory diseases and infections. Open Biology. 2023;**13**(3):230014

[115] Agus A et al. Western diet induces a shift in microbiota composition enhancing susceptibility to adherent-invasive *E. coli* infection and intestinal inflammation. Scientific Reports. 2016;**6**:19032

[116] Liu P et al. The role of short-chain fatty acids in intestinal barrier function, inflammation, oxidative stress, and colonic carcinogenesis. Pharmacological Research. 2021;**165**:105420

[117] Sinha AK et al. Dietary fibre directs microbial tryptophan metabolism via metabolic interactions in the gut microbiota. Nature Microbiology. 2024;**9**(8):1964-1978

[118] Tennoune N, Andriamihaja M, Blachier F. Production of indole and indole-related compounds by the intestinal microbiota and consequences for the host: The good, the bad, and the ugly. Microorganisms. 2022;**10**(5)

[119] Gao K et al. Tryptophan metabolism: A link between the gut microbiota and brain. Advances in Nutrition. 2020;**11**(3):709-723

[120] Williams BB et al. Discovery and characterization of gut microbiota decarboxylases that can produce the neurotransmitter tryptamine. Cell Host & Microbe. 2014;**16**(4):495-503

[121] Dodd D et al. A gut bacterial pathway metabolizes aromatic amino acids into nine circulating metabolites. Nature. 2017;**551**(7682):648-652

[122] Wlodarska M et al. Indoleacrylic acid produced by commensal *Peptostreptococcus* species suppresses inflammation. Cell Host & Microbe. 2017;**22**(1):25-37 e6

[123] Cervantes-Barragan L et al. *Lactobacillus reuteri* induces gut intraepithelial CD4(+)CD8alphaalpha(+) T cells. Science. 2017;**357**(6353):806-810

[124] Russell WR et al. Major phenylpropanoid-derived metabolites in the human gut can arise from microbial fermentation of protein. Molecular Nutrition & Food Research. 2013;**57**(3):523-535

[125] Rigottier-Gois L. Dysbiosis in inflammatory bowel diseases: The oxygen hypothesis. The ISME Journal. 2013;7(7):1256-1261

[126] Tang J et al. Biosynthetic pathways and functions of indole-3-acetic acid in microorganisms. Microorganisms. 2023;**11**(8)

[127] Kurnasov O et al. Aerobic tryptophan degradation pathway in bacteria: Novel kynurenine formamidase. FEMS Microbiology Letters. 2003;**227**(2):219-227

[128] Bortolotti P et al. Tryptophan catabolism in *Pseudomonas aeruginosa* and potential for inter-kingdom relationship. BMC Microbiology. 2016;**16**(1):137

[129] Chen Y, Guillemin GJ. Kynurenine pathway metabolites in humans: Disease and healthy states. International Journal of Tryptophan Research. 2009;**2**:1-19

[130] Ou W et al. The kynurenine pathway in major depressive disorder under different disease states: A systematic review and meta-analysis. Journal of Affective Disorders. 2023;**339**:624-632

[131] Lin P et al. Dysbiosis of the gut microbiota and kynurenine (Kyn) pathway activity as potential biomarkers in patients with major depressive disorder. Nutrients. 2023;**15**(7)

[132] Almulla AF et al. The tryptophan catabolite or kynurenine pathway in major depressive and bipolar disorder: A systematic review and meta-analysis. Brain, Behavior, & Immunity - Health. 2022;**26**:100537

[133] Arnone D et al. Role of kynurenine pathway and its metabolites in mood disorders: A systematic review and meta-analysis of clinical studies. Neuroscience and Biobehavioral Reviews. 2018;**92**:477-485

[134] Vints WAJ et al. Inflammatory blood biomarker kynurenine is linked with elevated neuroinflammation and neurodegeneration in older adults: Evidence from two 1H-MRS post-processing analysis methods. Frontiers in Psychiatry. 2022;**13**:859772

[135] Sorgdrager FJH et al. Age- and disease-specific changes of the kynurenine pathway in Parkinson's and Alzheimer's disease. Journal of Neurochemistry. 2019;**151**(5):656-668

[136] Apalset EM et al. Interferon (IFN)-gamma-mediated inflammation and the kynurenine pathway in relation to bone mineral density: The Hordaland Health Study. Clinical and Experimental Immunology. 2014;**176**(3):452-460

[137] Witoszynska-Sobkowiak J et al. The kynurenine pathway in patients with rheumatoid arthritis during tumor necrosis factor alpha inhibitors treatment. Reumatologia. 2024;**62**(4):220-225

[138] Chen LM et al. Tryptophan-kynurenine metabolism: A link between the gut and brain for depression in inflammatory bowel disease. Journal of Neuroinflammation. 2021;**18**(1):135

[139] Damerell V et al. Circulating tryptophan-kynurenine pathway metabolites are associated with all-cause mortality among patients with stage I-III colorectal cancer. International Journal of Cancer. 2025;**156**(3):552-565

[140] Girithar HN et al. Involvement of the kynurenine pathway in breast cancer: Updates on clinical research and trials. British Journal of Cancer. 2023;**129**(2):185-203

[141] Gouasmi R et al. The kynurenine pathway and cancer: Why keep it simple when you can make it complicated. Cancers (Basel). 2022;**14**(11)

[142] Leon-Letelier RA et al. The kynurenine pathway presents multi-faceted metabolic vulnerabilities in cancer. Frontiers in Oncology. 2023;**13**:1256769

[143] Yang Y et al. The role of the kynurenine pathway in cardiovascular disease. Frontiers in Cardiovascular Medicine. 2024;**11**:1406856

[144] Zaborin A et al. Spatial compartmentalization of the microbiome between the lumen and crypts is lost in the murine cecum following the process of surgery, including overnight fasting and exposure to antibiotics. mSystems. 2020;**5**(3)

[145] Loughman JA et al. Local generation of kynurenines mediates inhibition of neutrophil chemotaxis by *Uropathogenic Escherichia coli*. Infection and Immunity. 2016;**84**(4):1176-1183

[146] Mezrich JD et al. An interaction between kynurenine and the aryl hydrocarbon receptor can generate regulatory T cells. Journal of Immunology. 2010;**185**(6):3190-3198

[147] Sun J et al. Microbiota-derived metabolite indoles induced aryl hydrocarbon receptor activation and inhibited neuroinflammation in APP/PS1 mice. Brain, Behavior, and Immunity. 2022;**106**:76-88

[148] Powell DN et al. Indoles from the commensal microbiota act via the AHR and IL-10 to tune the cellular composition of the colonic epithelium during aging. Proceedings of the National Academy of Sciences of the United States of America. 2020;**117**(35):21519-21526

[149] Hezaveh K et al. Tryptophan-derived microbial metabolites activate the aryl hydrocarbon receptor in tumor-associated macrophages to suppress anti-tumor immunity. Immunity. 2022;**55**(2):324-340 e8

[150] Di Bella S et al. *Clostridium difficile* toxins A and B: Insights into pathogenic properties and extraintestinal effects. Toxins (Basel). 2016;**8**(5)

[151] Tucker KD, Carrig PE, Wilkins TD. Toxin A of *Clostridium difficile* is a potent cytotoxin. Journal of Clinical Microbiology. 1990;**28**(5):869-871

[152] Otto M. *Staphylococcus aureus* toxins. Current Opinion in Microbiology. 2014;**17**:32-37

[153] Pinchuk IV, Beswick EJ, Reyes VE. Staphylococcal enterotoxins. Toxins (Basel). 2010;**2**(8):2177-2197

[154] Peng L et al. Toxic shock syndrome toxin 1 induces immune response via the activation of NLRP3 inflammasome. Toxins (Basel). 2021;**13**(1)

[155] Elsherif HM et al. Staphylococcal enterotoxins and toxic shock syndrome Toxin-1 and their association among bacteremic and infective endocarditis patients in Egypt. BioMed Research International. 2020;**2020**:6981095

[156] Hunt JM. Shiga toxin-producing *Escherichia coli* (STEC). Clinics in Laboratory Medicine. 2010;**30**(1):21-45

[157] Sheikh A et al. Enterotoxigenic *Escherichia coli* heat-labile toxin drives enteropathic changes in small intestinal

epithelia. Nature Communications. 2022;**13**(1):6886

[158] Michalska M, Wolf P. Pseudomonas Exotoxin A: Optimized by evolution for effective killing. Frontiers in Microbiology. 2015;**6**:963

[159] Llanos A et al. Higher levels of *Pseudomonas aeruginosa* LasB elastase expression are associated with early-stage infection in cystic fibrosis patients. Scientific Reports. 2023;**13**(1):14208

[160] Nomura K et al. *Pseudomonas aeruginosa* elastase causes transient disruption of tight junctions and downregulation of PAR-2 in human nasal epithelial cells. Respiratory Research. 2014;**15**(1):21

[161] Hall S et al. Cellular effects of Pyocyanin, a secreted virulence factor of *Pseudomonas aeruginosa*. Toxins (Basel). 2016;**8**(8)

[162] Ran H, Hassett DJ, Lau GW. Human targets of *Pseudomonas aeruginosa* pyocyanin. Proceedings of the National Academy of Sciences of the United States of America. 2003;**100**(24):14315-14320

[163] Allen L et al. Pyocyanin production by *Pseudomonas aeruginosa* induces neutrophil apoptosis and impairs neutrophil-mediated host defenses in vivo. Journal of Immunology. 2005;**174**(6):3643-3649

[164] Coburn PS et al. Enterococcus faecalis senses target cells and in response expresses cytolysin. Science. 2004;**306**(5705):2270-2272

[165] Coburn PS, Gilmore MS. The *Enterococcus faecalis* cytolysin: A novel toxin active against eukaryotic and prokaryotic cells. Cellular Microbiology. 2003;**5**(10):661-669

[166] Silva GN et al. Neuroinflammatory approach to surgical trauma: Biomarkers and mechanisms of immune and neuroendocrine responses. Journal of Personalized Medicine. 2024;**14**(8)

[167] Haroutounian S. Postoperative opioids, endocrine changes, and immunosuppression. Pain Reports. 2018;**3**(2):e640

[168] Du W. Interactions between endogenous opioids and the immune system. Advances in Neurobiology. 2024;**35**:27-43

[169] Rutherford ST, Bassler BL. Bacterial quorum sensing: Its role in virulence and possibilities for its control. Cold Spring Harbor Perspectives in Medicine. 2012;**2**(11)

[170] Alotiby A. Immunology of stress: A review article. Journal of Clinical Medicine. 2024;**13**(21)

[171] Wu L et al. Recognition of host immune activation by *Pseudomonas aeruginosa*. Science. 2005;**309**(5735):774-777

[172] Zaborina O et al. Dynorphin activates quorum sensing quinolone signaling in *Pseudomonas aeruginosa*. PLoS Pathogens. 2007;**3**(3):e35

[173] Patel NJ et al. Recognition of intestinal epithelial HIF-1alpha activation by *Pseudomonas aeruginosa*. American Journal of Physiology. Gastrointestinal and Liver Physiology. 2007;**292**(1):G134-G142

[174] Zaborin A et al. Red death in *Caenorhabditis elegans* caused by *Pseudomonas aeruginosa* PAO1. Proceedings of the National Academy of Sciences of the United States of America. 2009;**106**(15):6327-6332

[175] Wong RS, Wirtz RA, Hancock RE. *Pseudomonas aeruginosa* outer membrane

protein OprF as an expression vector for foreign epitopes: The effects of positioning and length on the antigenicity of the epitope. Gene. 1995;**158**(1):55-60

[176] Fito-Boncompte L et al. Full virulence of *Pseudomonas aeruginosa* requires OprF. Infection and Immunity. 2011;**79**(3):1176-1186

[177] Fleming MA 2nd et al. The enteric nervous system and its emerging role as a therapeutic target. Gastroenterology Research and Practice. 2020;**2020**:8024171

[178] Shenoy SS, Lui F. Biochemistry, endogenous opioids. In: StatPearls. Treasure Island (FL); 2025

[179] Pathan H, Williams J. Basic opioid pharmacology: An update. British Journal of Pain. 2012;**6**(1):11-16

[180] Sobczak M et al. Physiology, signaling, and pharmacology of opioid receptors and their ligands in the gastrointestinal tract: Current concepts and future perspectives. Journal of Gastroenterology. 2014;**49**(1):24-45

[181] Szczepaniak A et al. The role of kappa opioid receptors in immune system - An overview. European Journal of Pharmacology. 2022;**933**:175214

[182] Martyniak A, Wedrychowicz A, Tomasik PJ. Endogenous opioids in crohn's disease. Biomedicines. 2023;**11**:7

[183] Babrowski T et al. *Pseudomonas aeruginosa* virulence expression is directly activated by morphine and is capable of causing lethal gut-derived sepsis in mice during chronic morphine administration. Annals of Surgery. 2012;**255**(2):386-393

[184] Herman TF, Cascella M, Muzio MR. Mu receptors. In: StatPearls. Treasure Island (FL); 2025

[185] Brogden KA et al. The nervous system and innate immunity: The neuropeptide connection. Nature Immunology. 2005;**6**(6):558-564

[186] Hughes DT et al. The QseC adrenergic signaling cascade in *Enterohemorrhagic E. coli* (EHEC). PLoS Pathogens. 2009;**5**(8):e1000553

[187] Boukerb AM et al. Inter-kingdom signaling of stress hormones: Sensing, transport and modulation of bacterial physiology. Frontiers in Microbiology. 2021;**12**:690942

[188] Laskowski MA, Kazmierczak BI. Mutational analysis of RetS, an unusual sensor kinase-response regulator hybrid required for *Pseudomonas aeruginosa* virulence. Infection and Immunity. 2006;**74**(8):4462-4473

[189] Zolfaghar I et al. Mutation of retS, encoding a putative hybrid two-component regulatory protein in *Pseudomonas aeruginosa*, attenuates multiple virulence mechanisms. Microbes and Infection. 2005;**7**(13):1305-1316

[190] Du X et al. Norepinephrine may promote the progression of *Fusobacterium nucleatum* related colorectal cancer via quorum sensing signalling. Virulence. 2024;**15**(1):2350904

[191] Lucca V et al. Influence of the norepinephrine and medium acidification in the growth and adhesion of *Salmonella* Heidelberg isolated from poultry. Microbial Pathogenesis. 2020;**138**:103799

[192] Sturbelle RT et al. The role of quorum sensing in *Escherichia coli* (ETEC) virulence factors. Veterinary Microbiology. 2015;**180**(3-4):245-252

[193] Freestone PP et al. Stimulation of bacterial growth by heat-stable, norepinephrine-induced autoinducers.

FEMS Microbiology Letters. 1999;**172**(1):53-60

[194] Brosnahan AJ et al. Norepinephrine potentiates proinflammatory responses of human vaginal epithelial cells. Journal of Neuroimmunology. 2013;**259**(1-2):8-16

[195] Eltzschig HK, Carmeliet P. Hypoxia and inflammation. The New England Journal of Medicine. 2011;**364**(7):656-665

[196] Bhandari T, Nizet V. Hypoxia-inducible factor (HIF) as a pharmacological target for prevention and treatment of infectious diseases. Infectious Disease and Therapy. 2014;**3**(2):159-174

[197] Schaffer K, Taylor CT. The impact of hypoxia on bacterial infection. The FEBS Journal. 2015;**282**(12):2260-2266

[198] Kohler JE et al. Components of intestinal epithelial hypoxia activate the virulence circuitry of *Pseudomonas*. American Journal of Physiology. Gastrointestinal and Liver Physiology. 2005;**288**(5):G1048-G1054

[199] Weimann A et al. ESPEN practical guideline: Clinical nutrition in surgery. Clinical Nutrition. 2021;**40**(7):4745-4761

[200] Krezalek MA et al. Influence of nutrition therapy on the intestinal microbiome. Current Opinion in Clinical Nutrition and Metabolic Care. 2017;**20**(2):131-137

[201] Long J et al. Depletion of intestinal phosphate after operative injury activates the virulence of *P. aeruginosa* causing lethal gut-derived sepsis. Surgery. 2008;**144**(2):189-197

[202] Malone M et al. Management of iron deficiency and anemia after Roux-en-Y gastric bypass surgery: An observational study. Surgery for Obesity and Related Diseases. 2013;**9**(6):969-974

[203] Corwin HL et al. Management of perioperative iron deficiency in cardiac surgery: A modified RAND delphi study. The Annals of Thoracic Surgery. 2022;**113**(1):316-323

[204] Yilmaz B, Li H. Gut microbiota and iron: The crucial actors in health and disease. Pharmaceuticals (Basel). 2018;**11**(4)

[205] Abbaspour N, Hurrell R, Kelishadi R. Review on iron and its importance for human health. Journal of Research in Medical Sciences : The Official Journal of Isfahan University of Medical Sciences. 2014;**19**(2):164-174

[206] Ullah I, Lang M. Key players in the regulation of iron homeostasis at the host-pathogen interface. Frontiers in Immunology. 2023;**14**:1279826

[207] Fleming ID et al. Modeling Acinetobacter baumannii wound infections: The critical role of iron. Journal of Trauma and Acute Care Surgery. 2017;**82**(3):557-565

[208] Piskin E et al. Iron absorption: Factors, limitations, and improvement methods. ACS Omega. 2022;**7**(24):20441-20456

[209] Kim M et al. *Pseudomonas aeruginosa* wound infection involves activation of its iron acquisition system in response to fascial contact. Journal of Trauma and Acute Care Surgery. 2015;**78**(4):823-829

[210] Sangkhae V, Nemeth E. Regulation of the iron homeostatic hormone Hepcidin. Advances in Nutrition. 2017;**8**(1):126-136

[211] Ganz T. Hepcidin-a regulator of intestinal iron absorption and iron

recycling by macrophages. Best Practice & Research. Clinical Haematology. 2005;**18**(2):171-182

[212] Nemeth E, Ganz T. Hepcidin-Ferroportin interaction controls systemic iron homeostasis. International Journal of Molecular Sciences. 2021;**22**(12)

[213] Kell DB, Heyden EL, Pretorius E. The biology of Lactoferrin, an iron-binding protein that can help defend against viruses and bacteria. Frontiers in Immunology. 2020;**11**:1221

[214] Wilson BR et al. Siderophores in iron metabolism: From mechanism to therapy potential. Trends in Molecular Medicine. 2016;**22**(12):1077-1090

[215] Porcheron G, Dozois CM. Interplay between iron homeostasis and virulence: Fur and RyhB as major regulators of bacterial pathogenicity. Veterinary Microbiology. 2015;**179**(1-2):2-14

[216] Lamont IL et al. Siderophore-mediated signaling regulates virulence factor production in *Pseudomonas aeruginosa*. Proceedings of the National Academy of Sciences of the United States of America. 2002;**99**(10):7072-7077

[217] Fiedler HP et al. Enterobactin: The characteristic catecholate siderophore of *Enterobacteriaceae* is produced by *Streptomyces* species.(1). FEMS Microbiology Letters. 2001;**196**(2):147-151

[218] Hammer ND, Skaar EP. Molecular mechanisms of *Staphylococcus aureus* iron acquisition. Annual Review of Microbiology. 2011;**65**:129-147

[219] Griffiths GL et al. Vibriobactin, a siderophore from vibrio cholerae. The Journal of Biological Chemistry. 1984;**259**(1):383-385

[220] Chakraborty K et al. Bacillibactin class of siderophore antibiotics from a marine symbiotic bacillus as promising antibacterial agents. Applied Microbiology and Biotechnology. 2022;**106**(1):329-340

[221] Lee JY et al. Regulation of petrobactin and bacillibactin biosynthesis in bacillus anthracis under iron and oxygen variation. PLoS One. 2011;**6**(6):e20777

[222] Reddy PV et al. Disruption of mycobactin biosynthesis leads to attenuation of *Mycobacterium tuberculosis* for growth and virulence. The Journal of Infectious Diseases. 2013;**208**(8):1255-1265

[223] Weakland DR et al. The *Serratia marcescens* Siderophore Serratiochelin is necessary for full virulence during bloodstream infection. Infection and Immunity. 2020;**88**(8)

[224] Andrews SC, Robinson AK, Rodriguez-Quinones F. Bacterial iron homeostasis. FEMS Microbiology Reviews. 2003;**27**(2-3):215-237

[225] Troxell B, Hassan HM. Transcriptional regulation by ferric uptake regulator (Fur) in pathogenic bacteria. Frontiers in Cellular and Infection Microbiology. 2013;**3**:59

[226] Bradley JM et al. Bacterial iron detoxification at the molecular level. The Journal of Biological Chemistry. 2020;**295**(51):17602-17623

[227] Page MGP. The role of iron and siderophores in infection, and the development of siderophore antibiotics. Clinical Infectious Diseases. 2019;**69**(Suppl. 7):S529-S537

[228] Abbas M et al. Effects of iron deficiency and iron supplementation at the host-microbiota interface: Could a piglet model unravel complexities of the

underlying mechanisms? Frontiers in Nutrition. 2022;**9**:927754

[229] Carvalho C et al. Hepcidin and disordered mineral metabolism in chronic kidney disease. Clinical Nephrology. 2011;**76**(2):90-98

[230] Erben RG. Physiological actions of fibroblast growth Factor-23. Frontiers in Endocrinology (Lausanne). 2018;**9**:267

[231] Agoro R, White KE. Regulation of FGF23 production and phosphate metabolism by bone-kidney interactions. Nature Reviews. Nephrology. 2023;**19**(3):185-193

[232] Seweryn P et al. Structural insights into the bacterial carbon-phosphorus lyase machinery. Nature. 2015;**525**(7567):68-72

[233] Zimmermann H, Zebisch M, Strater N. Cellular function and molecular structure of ecto-nucleotidases. Purinergic Signal. 2012;**8**(3):437-502

[234] Santos-Beneit F. The Pho regulon: A huge regulatory network in bacteria. Frontiers in Microbiology. 2015;**6**:402

[235] Lamarche MG et al. The phosphate regulon and bacterial virulence: A regulatory network connecting phosphate homeostasis and pathogenesis. FEMS Microbiology Reviews. 2008;**32**(3):461-473

[236] Zaborina O et al. Structure-function aspects of PstS in multi-drug-resistant *Pseudomonas aeruginosa*. PLoS Pathogens. 2008;**4**(2):e43

[237] Zaborina O et al. Microscopic analysis: Morphotypes and cellular appendages. Methods in Molecular Biology. 2014;**1149**:99-107

[238] Chekabab SM, Harel J, Dozois CM. Interplay between genetic regulation of phosphate homeostasis and bacterial virulence. Virulence. 2014;**5**(8):786-793

[239] Borody TJ et al. Bacteriotherapy using fecal flora: Toying with human motions. Journal of Clinical Gastroenterology. 2004;**38**(6):475-483

[240] Bornbusch SL et al. Fecal microbiota transplants facilitate post-antibiotic recovery of gut microbiota in cheetahs (*Acinonyx jubatus*). Communications Biology. 2024;**7**(1):1689

[241] Fillon M. Fecal microbiota transplants may aid melanoma immunotherapy resistance. CA: A Cancer Journal for Clinicians. 2021;**71**(4):285-286

[242] He Z et al. Multiple fresh fecal microbiota transplants induces and maintains clinical remission in Crohn's disease complicated with inflammatory mass. Scientific Reports. 2017;**7**(1):4753

[243] Khan R et al. Fecal microbiota transplants for inflammatory bowel disease treatment: Synthetic- and engineered communities-based microbiota transplants are the future. Gastroenterology Research and Practice. 2022;**2022**:9999925

[244] Mamoon L, Olesen SW. Fecal microbiota transplants annually and their positive clinical impact. Clinical and Translational Gastroenterology. 2020;**11**(11):e00247

[245] Utay NS et al. Evaluation of six weekly oral fecal microbiota transplants in people with HIV. Pathogens & Immunity. 2020;**5**(1):364-381

[246] Baktash A et al. Mechanistic insights in the success of fecal microbiota transplants for the treatment of *Clostridium difficile* infections. Frontiers in Microbiology. 2018;**9**:1242

[247] Lin SC, Alonso CD, Moss AC. Fecal microbiota transplantation for recurrent *Clostridium difficile* infection in patients with solid organ transplants: An institutional experience and review of the literature. Transplant Infectious Disease. 2018;**20**(6):e12967

[248] Walton J, Burns D, Gaehle KE. Process and outcome of fecal microbiota transplants in patients with recurrent *Clostridium difficile* infection: A prospective study. Gastroenterology Nursing. 2017;**40**(5):411-419

[249] Kim SM et al. Fecal microbiota transplant rescues mice from human pathogen mediated sepsis by restoring systemic immunity. Nature Communications. 2020;**11**(1):2354

[250] Zaborin A et al. Critical role of microbiota within cecal crypts on the regenerative capacity of the intestinal epithelium following surgical stress. American Journal of Physiology. Gastrointestinal and Liver Physiology. 2017;**312**(2):G112-G122

[251] Vendrik KEW et al. Transmission of antibiotic-susceptible *Escherichia coli* causing urinary tract infections in a fecal microbiota transplantation recipient: Consequences for donor screening? Open Forum Infectious Diseases. 2022;**9**(7):ofac324

[252] Keskey R et al. Defining microbiome readiness for surgery: Dietary prehabilitation and stool biomarkers as predictive tools to improve outcome. Annals of Surgery. 2022;**276**(5):e361-e369

[253] West MA, Wischmeyer PE, Grocott MPW. Prehabilitation and nutritional support to improve perioperative outcomes. Current Anesthesiology Reports. 2017;7(4):340-349

[254] Drevik J, Michel C, Hamilton-Reeves J. Nutritional prehabilitation: A pragmatic guide. European Urology Focus. 2024;**10**(1):11-12

[255] Vigneswaran J et al. Western diet-induced transcriptional changes in anastomotic tissue is associated with early local recurrence in a mouse model of colorectal surgery. Annals of Surgery. 2023;**278**(6):954-960

[256] Gaines S et al. Western diet promotes intestinal colonization by Collagenolytic microbes and promotes tumor formation after colorectal surgery. Gastroenterology. 2020;**158**(4):958-970 e2

[257] Hyoju SK et al. Mice fed an obesogenic Western diet, administered antibiotics, and subjected to a sterile surgical procedure develop lethal Septicemia with multidrug-resistant Pathobionts. MBio. 2019;**10**(4)

[258] Severino A et al. The microbiome-driven impact of western diet in the development of noncommunicable chronic disorders. Best Practice & Research. Clinical Gastroenterology. 2024;**72**:101923

[259] Falcone L et al. Nutritional prehabilitation intervention in hematological patients undergoing bone marrow transplant: A systematic review of the literature. Nutrients. 2024;**16**(24)

[260] Sole-Sedeno JM et al. Protein supplementation in a prehabilitation program in patients undergoing surgery for endometrial cancer. International Journal of Environmental Research and Public Health. 2023;**20**(8)

[261] Suenaga H et al. Impact of dental referral prior to elective surgery on postoperative outcomes. Journal of the American Medical Directors Association. 2022;**23**(4):638-645 e2

[262] Ramos Meyers G, Samouda H, Bohn T. Short chain fatty acid metabolism in relation to gut microbiota and genetic variability. Nutrients. 2022;**14**(24)

[263] Niccolai E et al. Evaluation and comparison of short chain fatty acids composition in gut diseases. World Journal of Gastroenterology. 2019;**25**(36):5543-5558

[264] Sanna S et al. Causal relationships among the gut microbiome, short-chain fatty acids and metabolic diseases. Nature Genetics. 2019;**51**(4):600-605

[265] Hendrikx T, Schnabl B. Indoles: Metabolites produced by intestinal bacteria capable of controlling liver disease manifestation. Journal of Internal Medicine. 2019;**286**(1):32-40

[266] Hodgkinson K et al. Butyrate's role in human health and the current progress towards its clinical application to treat gastrointestinal disease. Clinical Nutrition. 2023;**42**(2):61-75

[267] Chen J, Vitetta L. The role of butyrate in attenuating Pathobiont-induced hyperinflammation. Immune Network. 2020;**20**(2):e15

[268] Sekhavat A, Sun JM, Davie JR. Competitive inhibition of histone deacetylase activity by trichostatin A and butyrate. Biochemistry and Cell Biology. 2007;**85**(6):751-758

[269] Davie JR. Inhibition of histone deacetylase activity by butyrate. The Journal of Nutrition. 2003;**133**(7 Suppl.):2485S-2493S

[270] Pieniawska M, Izykowska K. Role of histone deacetylases in T-cell development and function. International Journal of Molecular Sciences. 2022;**23**(14)

[271] Peng L et al. Butyrate enhances the intestinal barrier by facilitating tight junction assembly via activation of AMP-activated protein kinase in Caco-2 cell monolayers. The Journal of Nutrition. 2009;**139**(9):1619-1625

[272] Di Vincenzo F et al. Gut microbiota, intestinal permeability, and systemic inflammation: A narrative review. Internal and Emergency Medicine. 2024;**19**(2):275-293

[273] Liu H et al. Butyrate: A double-edged sword for health? Advances in Nutrition. 2018;**9**(1):21-29

[274] Klojdova I et al. Encapsulation: A strategy to deliver therapeutics and bioactive compounds? Pharmaceuticals (Basel). 2023;**16**(3)

[275] Kennedy PJ et al. Kynurenine pathway metabolism and the microbiota-gut-brain axis. Neuropharmacology. 2017;**112**(Pt. B):399-412

[276] Saraf MK et al. Formula diet driven microbiota shifts tryptophan metabolism from serotonin to tryptamine in neonatal porcine colon. Microbiome. 2017;**5**(1):77

[277] Scott SA, Fu J, Chang PV. Microbial tryptophan metabolites regulate gut barrier function via the aryl hydrocarbon receptor. Proceedings of the National Academy of Sciences of the United States of America. 2020;**117**(32):19376-19387

[278] Assimakopoulos SF et al. Gut-origin sepsis in the critically ill patient: Pathophysiology and treatment. Infection. 2018;**46**(6):751-760

[279] Deitch EA. Gut-origin sepsis: Evolution of a concept. The Surgeon. 2012;**10**(6):350-356

[280] Alverdy J et al. Gut-derived sepsis occurs when the right pathogen with

the right virulence genes meets the right host: Evidence for in vivo virulence expression in *Pseudomonas aeruginosa*. Annals of Surgery. 2000;**232**(4):480-489

[281] Alverdy JC, Laughlin RS, Wu L. Influence of the critically ill state on host-pathogen interactions within the intestine: Gut-derived sepsis redefined. Critical Care Medicine. 2003;**31**(2):598-607

[282] Oami T et al. Claudin-2 upregulation enhances intestinal permeability, immune activation, dysbiosis, and mortality in sepsis. Proceedings of the National Academy of Sciences of the United States of America. 2024;**121**(10):e2217877121

[283] Zhang G et al. Artificial mucus layer formed in response to ROS for the oral treatment of inflammatory bowel disease. Science Advances. 2024;**10**(30):eado8222

[284] Detwiler RE, Kramer JR. Preparation and applications of artificial mucins in biomedicine. Current Opinion in Solid State & Materials Science. 2022;**26**(6)

[285] Wynne AC et al. Experimental and modeling approaches to determine drug diffusion coefficients in artificial mucus. Heliyon. 2024;**10**(20):e38638

[286] Wu L et al. High-molecular-weight polyethylene glycol prevents lethal sepsis due to intestinal *Pseudomonas aeruginosa*. Gastroenterology. 2004;**126**(2):488-498

[287] Valuckaite V et al. Oral PEG 15-20 protects the intestine against radiation: Role of lipid rafts. American Journal of Physiology. Gastrointestinal and Liver Physiology. 2009;**297**(6):G1041-G1052

[288] Mao J et al. De novo synthesis of phosphorylated triblock copolymers with pathogen virulence-suppressing properties that prevent infection-related mortality. ACS Biomaterials Science & Engineering. 2017;**3**(9):2076-2085

[289] Edelstein A et al. Protective effects of nonionic triblock copolymers on bile acid-mediated epithelial barrier disruption. Shock. 2011;**36**(5):451-457

[290] Olivas AD et al. Intestinal tissues induce an SNP mutation in *Pseudomonas aeruginosa* that enhances its virulence: Possible role in anastomotic leak. PLoS One. 2012;**7**(8):e44326

[291] Valuckaite V et al. High molecular weight polyethylene glycol (PEG 15-20) maintains mucosal microbial barrier function during intestinal graft preservation. The Journal of Surgical Research. 2013;**183**(2):869-875

[292] Zaborin A et al. *Pseudomonas aeruginosa* overrides the virulence inducing effect of opioids when it senses an abundance of phosphate. PLoS One. 2012;**7**(4):e34883

[293] Romanowski K et al. Prevention of siderophore- mediated gut-derived sepsis due to P. Aeruginosa can be achieved without iron provision by maintaining local phosphate abundance: Role of pH. BMC Microbiology. 2011;**11**:212

[294] Wiegerinck M et al. Novel de novo synthesized phosphate carrier compound ABA-PEG20k-Pi20 suppresses collagenase production in enterococcus faecalis and prevents colonic anastomotic leak in an experimental model. The British Journal of Surgery. 2018;**105**(10):1368-1376

[295] Hyoju SK et al. A novel nonantibiotic gut-directed strategy to prevent surgical site infections. Annals of Surgery. 2022;**276**(3):472-481

Section 2

Gut Microbiota Disorders and Neurodegenerative Diseases

Chapter 3

Gut Microbiota Modulation in Parkinson's Disease: Exploring Phytotherapy, Pharmacotherapy, and Precision Interventions

Samuel Oluwadare Olalekan, Patrick Godwin Okwute, Ifabunmi Oduyemi Osonuga, Adekunle Muiz Mofolorunso, Bukunola Oluyemisi Adegbesan, Esther Nkechi Ezima and Gideon Olamilekan Oluwatunase

Abstract

Parkinson's disease (PD) is a progressive neurodegenerative disorder with growing evidence implicating gut microbiota dysbiosis in its pathogenesis. Recent research highlights the gut-brain axis as a crucial therapeutic target, with interventions such as phytotherapy, pharmacotherapy, and precision medicine emerging as promising approaches. Phytotherapy, including plant-derived compounds like curcumin and dioscin, has demonstrated neuroprotective effects by modulating gut microbiota, reducing neuroinflammation, and enhancing short-chain fatty acid production. Pharmacotherapy, particularly levodopa, remains the primary treatment for PD motor symptoms, but its interactions with gut microbiota can affect drug metabolism, necessitating microbiota-targeted strategies such as enzyme inhibitors and antibiotic co-administration. Precision interventions, including probiotics, prebiotics, synbiotics, dietary modifications, and fecal microbiota transplantation, aim to restore microbial balance and improve gut-brain axis function. This chapter explores the mechanisms, clinical outcomes, and limitations of these interventions, emphasizing the need for integrative and personalized therapeutic approaches. By targeting gut microbiota, these strategies hold potential for optimizing PD management, enhancing treatment efficacy, and improving patient quality of life.

Keywords: Parkinson's disease, gut microbiota, microbiota-gut-brain axis, phytotherapy, pharmacotherapy, precision medicine, levodopa, probiotics, synbiotics, fecal microbiota transplantation

1. Introduction

Parkinson's disease (PD) is a prevalent chronic neurodegenerative disorder resulting from significant damage to dopamine-producing neurons, leading to dopamine deficits in the midbrain and subsequent alterations in various neurotransmitters, including glutamate, GABA, and serotonin [1]. Globally, disability and mortality rates associated with PD have surged, surpassing those of other neurological disorders. In 2019 alone, PD caused 5.8 million disability-adjusted life years and 329,000 deaths, marking an 81% increase in disability and over 100% rise in mortality since 2000 [2, 3]. This neural degeneration manifests not only in typical motor symptoms such as resting tremor, bradykinesia, and muscle stiffness but also encompasses a wide array of non-motor symptoms, including constipation, olfactory impairments, sleep disturbances, autonomic nervous dysfunction, and cognitive and psychiatric disorders [4].

Despite its clinical complexity, current therapeutic interventions for PD primarily focus on managing motor symptoms through dopaminergic medications. Current therapeutic strategies for PD focus primarily on managing motor symptoms through pharmacotherapy. Levodopa remains the cornerstone of PD treatment, serving as a dopamine precursor that replenishes striatal dopamine. However, its long-term use is associated with complications, including peripheral conversion-related complications and long-term adverse events [1]. The gut microbiota, a diverse community of microorganisms residing in the gastrointestinal tract, has been implicated in several diseases, including Parkinson's disease (PD), due to its critical role in immune regulation, metabolism, and the gut-brain axis [5–7].

Emerging research suggests that targeting the gut microbiota may open new therapeutic avenues for Parkinson's disease (PD) treatment [8–10]. Recent studies have highlighted the bidirectional relationship between gut microbiota and PD, emphasizing the potential of microbiome-targeted interventions to alleviate symptoms and modify disease progression [11, 12]. This chapter seeks to examine how phytotherapy, pharmacotherapy, and precision medicine target the gut microbiome to treat PD, focusing on their mechanisms of action, therapeutic potential, and clinical implications. By incorporating these interventions, researchers hope to revolutionize PD management by enhancing treatment efficacy, reducing complications, and addressing disease mechanisms at the microbiome level.

2. Pathophysiology of Parkinson's disease

Recent research has provided significant insights into the multifaceted pathophysiology of PD, underscoring its complex and progressive nature. Central to PD is the degeneration of dopamine-producing neurons in the substantia nigra pars compacta, a critical component of the basal ganglia. This neuronal loss leads to a marked reduction in dopamine levels, crucial for regulating motor functions, resulting in the hallmark motor symptoms of PD such as tremors, rigidity, and bradykinesia [13]. A defining pathological feature of PD is the presence of Lewy bodies and Lewy neurites, which are abnormal aggregates of the alpha-synuclein protein. These inclusions interfere with normal cellular functions and contribute significantly to neuronal death and disease progression [14].

Oxidative stress plays a crucial role in the pathophysiology of PD. It contributes to mitochondrial dysfunction and the formation of abnormal protein aggregates, which are central to PD progression. Studies have demonstrated that interventions like melatonin can mitigate oxidative stress, suggesting potential therapeutic benefits. For example, melatonin administration has been shown to significantly reduce oxidative stress markers and improve mitochondrial activity in PD patients, highlighting its neuroprotective effects [15]. Additionally, oxidative stress-related abnormalities, such as elevated plasma NfL levels, have been linked to brain volume loss and neuropsychiatric deficits in PD, underscoring the importance of targeting oxidative stress in therapeutic strategies [16].

Additionally, other novel aspects of PD pathophysiology have been explored, including cerebral microbleeds (CMB), axial postural abnormalities (APAs), and immune activation. CMBs are associated with hypertension in PD, distinct from the amyloid-β burden seen in Alzheimer's disease, suggesting different underlying mechanisms [17]. APAs, such as camptocormia and Pisa syndrome, significantly impair quality of life, with botulinum toxin treatments combined with rehabilitation showing promise in management [18]. Peripheral immune activation, evidenced by elevated proinflammatory cytokines, underscores the role of neuroinflammation in PD, although the specific contributions of cytokines like IL-17 remain under investigation [19]. These findings collectively underscore the intricate pathophysiology of PD, necessitating a multidisciplinary approach for effective diagnosis and treatment.

Beyond the traditional understanding, recent studies have shed light on additional neurodegenerative mechanisms and contributing factors. For instance, alpha-synuclein seed amplification assays (α-syn SAAs) have been developed as highly sensitive and specific diagnostic tools, capable of detecting misfolded alpha-synuclein in cerebrospinal fluid and other tissues, thus distinguishing PD from other neurodegenerative disorders [20]. Genetic mutations, particularly in genes such as SNCA, LRRK2, and PARK2, are implicated in familial forms of PD, while environmental factors, such as exposure to pesticides, are more commonly associated with sporadic cases [14]. Emerging evidence also points to the gut-brain axis playing a pivotal role in PD pathophysiology. Dysbiosis of the gut microbiota and resulting gut-brain axis disturbances are believed to contribute to disease progression, with interventions like fecal microbiota transplantation showing potential in modifying motor and non-motor symptoms [21, 22].

3. Exploring phytotherapy, pharmacotherapy, and precision interventions in Parkinson's disease management

The gut-brain axis, a bidirectional communication network between the central nervous system and the gastrointestinal tract, has emerged as a key factor in the pathology of Parkinson's disease (PD). Increasing evidence suggests that gut dysbiosis, neuroinflammation, and microbial metabolite imbalances play a critical role in disease progression. As a result, novel therapeutic strategies are being developed to target gut microbiota as a means of modulating PD pathology. For the purpose of this chapter, these strategies can be categorized into three primary approaches: phytotherapy, which utilizes plant-derived bioactive compounds; pharmacotherapy, which includes conventional drug-based treatments such as levodopa; and precision medicine, which focuses on personalized interventions such as dietary modifications, probiotics, and fecal microbiota transplantation (FMT). Each of these approaches offers unique mechanisms of action and therapeutic implications for PD management.

3.1 Phytotherapies in the management of Parkinson's disease

Plant-based therapies have gained significant attention for their potential to modulate gut microbiota, reduce neuroinflammation, and enhance neuroprotection in PD. Many plant-derived compounds, including polyphenols, alkaloids, and flavonoids, possess antioxidant and anti-inflammatory properties that may influence the gut-brain axis [23, 24]. Curcumin, for example, has been shown to enhance beneficial gut bacteria while reducing pathogenic species, leading to improved dopamine synthesis and decreased α-synuclein aggregation [25]. Also, given the critical role of oxidative stress and neuroinflammation in PD progression, there is a strong rationale for exploring botanical sources as therapeutic options. Botanical compounds, known for their antioxidant, anti-inflammatory, and neuroprotective properties, offer promising avenues for counteracting the mechanisms driving PD. As such, the integration of botanical treatments may not only alleviate symptoms but also target the disease's underlying causes, underscoring the potential for natural compounds to contribute significantly to PD management and therapy (**Table 1**).

Prebiotic supplementation has also demonstrated potential in influencing gut microbiota composition and promoting short-chain fatty acid (SCFA) production, particularly butyrate, which possesses anti-inflammatory properties. Hall et al. [26] showed that prebiotic fiber intervention increased SCFA levels, enhanced gut microbial diversity, and reduced neurofilament light chain (NfL), a marker of neurodegeneration, in PD patients. However, the study was limited by its short intervention period, making it difficult to assess long-term effects. A similar prebiotic-based strategy was explored by Liu et al. [32], who investigated polymannuronic acid (PM) in combination with *Lacticaseibacillus rhamnosus* GG probiotics. Their study

Treatment	Mechanism	Positive outcome(s)	Limitation(s)	References
Prebiotic fiber	SCFA production	Reduced inflammation, improved microbial diversity	Short study duration	[26]
Curcumin	Microbiota reprogramming, dopamine synthesis	Improved motor function, reduced α-synuclein aggregation	Bioavailability issues	[25]
Berberine	Tyrosine hydroxylase activation in gut bacteria	Increased dopamine levels, improved motor function	Microbial resistance concerns	[27]
Nardosinone	Gut microbiota modulation, levodopa bioavailability	Increased neurotransmitter levels, improved motor function	Human validation needed	[28]
Trehalose	Autophagy induction, gut microbiota restoration	Reduced α-synuclein aggregation, neuroprotection	Lacks human trials	[5]
Dioscin	Bile acid regulation, GLP-1 activation	Improved motor function, reduced neuroinflammation	No human studies	[29]
Perilla Seed Oil	SCFA enhancement, anti-inflammatory	Reduced neuroinflammation, improved gut diversity	Requires long-term adherence	[30]
Maslinic Acid	Microbiota modulation, neurotransmitter increase	Enhanced motor function	Dose-dependent effects unclear	[31]

Table 1.
Plant-based therapies in Parkinson's disease management: targeting gut microbiota.

demonstrated improvements in gut barrier integrity, enhanced SCFA-mediated anti-inflammatory effects, and increased levels of neurotrophic factors such as brain-derived neurotrophic factor (BDNF) and glial cell line-derived neurotrophic factor (GDNF). These findings highlight the potential of prebiotics and probiotics as neuroprotective agents, but further human studies are required to confirm their clinical efficacy.

Traditional herbal medicine, particularly formulations within Traditional Chinese Medicine (TCM), has also been investigated for its role in gut microbiota modulation and neuroprotection. Zhu et al. [33] reported that TCM formulations reduced intestinal permeability, mitigated gut dysbiosis, and enhanced dopamine metabolism, suggesting their potential to restore gut-brain axis homeostasis. However, the lack of standardization and regulatory oversight remains a major challenge in translating TCM therapies into clinical practice. Nardosinone, a bioactive compound derived from Nardostachys jatamansi, has also been studied for its potential to enhance levodopa bioavailability via gut microbiota modulation. Liu et al. [28] found that nardosinone increased beneficial gut microbial populations, leading to improved neurotransmitter levels (dopamine, serotonin) in the striatum, plasma, and feces of rotenone-induced PD rats. The compound also exhibited antioxidant and anti-inflammatory properties, reducing neuronal damage and improving motor function. However, its efficacy in human PD patients remains unverified, and the exact microbiota pathways involved require further investigation.

Additionally, dioscin, a bioactive steroidal saponin, has been shown to influence gut microbiota composition and bile acid metabolism in PD models. Mao et al. [29] found that dioscin reduced the Firmicutes-to-Bacteroidetes ratio and decreased bile salt hydrolase (BSH) activity, thereby increasing neuroprotective bile acids such as ursodeoxycholic acid and tauroursodeoxycholic acid. Activation of the glucagon-like peptide-1 (GLP-1) pathway further enhanced anti-inflammatory effects, contributing to improved motor function and reduced neuroinflammation. However, human studies are still needed to determine whether dioscin could serve as a viable PD treatment.

3.2 Pharmacotherapy in the management of Parkinson's disease

Pharmacotherapy remains the cornerstone of Parkinson's disease (PD) treatment, primarily aimed at dopamine replacement and symptomatic relief (**Table 2**). Levodopa, often co-administered with carbidopa to enhance bioavailability, remains the gold standard for managing motor symptoms. However, recent studies have revealed complex interactions between levodopa metabolism and gut microbiota. Certain bacterial species, such as *Enterococcus faecalis*, metabolize levodopa into dopamine before it reaches the brain, thereby reducing its therapeutic efficacy [34]. Strategies to optimize levodopa absorption, such as gut-directed antibiotics (e.g., vancomycin) and probiotic co-administration, have been explored, yet concerns about antibiotic resistance persist [39].

Beyond levodopa, other pharmacological approaches have been investigated for their effects on gut microbiota and neuroinflammation. Rifaximin and ceftriaxone, both antibiotics, have shown potential in modifying gut microbiota composition and reducing inflammation in PD models [37, 40]. However, while these interventions may offer short-term benefits, long-term microbiome alterations raise concerns about dysbiosis and microbial resistance. Another advancement, levodopa-carbidopa intestinal gel (LCIG), was introduced to bypass gastrointestinal metabolism and

Levodopa	**Gut microbial metabolism of levodopa**	**Increased dopamine levels, symptom relief**	**Gut dysbiosis, variability in response**	[34].
Levodopa + FMT	Microbiota restoration	Improved levodopa absorption	Individual variability, speculative impact	[35, 36]
Rifaximin	Microbiota modulation, inflammation reduction	Improved motor function, BBB integrity	Limited microbiota changes in human trials	[37]
Vancomycin	TLR4/NF-κB inhibition, dopamine metabolism	Reduced neuroinflammation, improved motor function	Risk of antibiotic resistance	[38]
Bacteriophage Therapy	Elimination of *E. faecalis*	Improved levodopa efficacy	Requires further validation	[8]

Table 2.
Pharmacotherapies in Parkinson's disease management: targeting gut microbiota.

provide continuous drug delivery. However, a study indicate that LCIG therapy alters gut microbial composition, increasing pro-inflammatory Proteobacteria, Enterobacteriaceae, and a reduction of Firmicutes, Lachnospiraceae, and Blautia, which may exacerbate gut inflammation over time [41].

3.2.1 Microbiota-targeted strategies to optimize levodopa therapy

Levodopa metabolism is also affected by specific bacterial enzymes. Cheng et al. [34] highlighted that *Enterococcus faecalis* metabolizes levodopa before it reaches the brain, reducing its availability for therapeutic effects. Their study identified Mito-ortho-HNK, a compound that inhibits *E. faecalis*, as a potential solution to enhance levodopa absorption and improve motor symptoms. While promising, interindividual differences in gut microbiota composition complicate the generalizability of these microbiota-targeted levodopa modifications, and their long-term consequences remain uncertain. Another critical factor influencing levodopa metabolism is the microbial enzyme tyrosine decarboxylase (tyrDC). Zhang et al. [42] identified that higher tyrDC gene abundance in *E. faecalis* correlated with reduced levodopa efficacy in PD patients. Their study emphasized that microbial metabolism of levodopa is a more significant factor in drug response than bile acid composition, as He et al. [43] found no significant association between fecal bile acids and levodopa metabolism.

3.2.2 Antibiotics in PD management: Potential and limitations

Beyond levodopa, antibiotic therapies have been explored for their potential in modifying gut microbiota and reducing neuroinflammation. Hong et al. [37] investigated the impact of rifaximin on gut microbiota modulation, demonstrating that it reduced Prevotellaceae while increasing Bacteroides, which led to lower systemic inflammation and enhanced blood-brain barrier (BBB) integrity. Their study also found that rifaximin improved motor and memory function in transgenic PD mice while decreasing pro-inflammatory cytokines such as IL-1β, IL-6, and TNF-α. However, the long-term effects of rifaximin on gut microbiota remain unclear, and human studies have shown limited microbiota changes after short-term treatment.

Baizabal-Carvallo et al. [44] evaluated intestinal decontamination therapy using rifaximin and polyethylene glycol, showing that this approach reduced levodopa-degrading bacteria and improved levodopa-induced dyskinesia and motor fluctuations. Although 80% of patients reported moderate to robust improvement, the study's small sample size and lack of long-term safety data require further validation.

Vancomycin has also been investigated for its effects on neuroinflammation and gut microbiota composition. Cui et al. [38] demonstrated that vancomycin suppressed inflammation through the TLR4/MyD88/NF-κB/TNF-α signaling pathway, while also increasing Akkermansia and Blautia levels, both associated with gut homeostasis and neuroprotection. Additionally, vancomycin modified dopamine metabolism by suppressing monoamine oxidase B (MAO-B) expression, leading to reduced dopamine degradation and improved motor function in MPTP-induced PD mice. Despite these benefits, concerns persist regarding vancomycin's long-term impact on gut microbiota stability and the risk of antibiotic resistance. Zhou et al. [40] further explored ceftriaxone's effects on gut inflammation and dopaminergic neuron survival, finding that it increased beneficial Akkermansia while reducing pathogenic Proteus. Their study suggests that ceftriaxone could serve as a potential PD-modifying therapy, but its long-term effects require further investigation.

3.3 Precision therapies in the management of Parkinson's disease

Precision medicine in Parkinson's disease (PD) aims to tailor interventions based on individual patient profiles, considering factors such as gut microbiota composition, metabolic markers, and genetic predisposition. Unlike conventional pharmacotherapy, which follows a standardized treatment model, precision medicine seeks to optimize therapeutic outcomes by addressing patient-specific variables. This approach includes dietary modifications, probiotics, synbiotics, and fecal microbiota transplantation (FMT) (**Table 3**).

3.3.1 Dietary interventions in Parkinson's disease management: Targeting gut microbiota

Dietary interventions, including the ketogenic diet (KD), Mediterranean diet (MediDiet), and low-protein, high-carbohydrate (LPHC) diets, have shown promise in modulating gut microbiota and alleviating PD symptoms. Jiang et al. [45, 47] demonstrated that KD reversed MPTP-induced gut dysbiosis by reducing pro-inflammatory Desulfovibrio while increasing beneficial Dubosiella, leading to improved motor function and reduced neuroinflammation. However, long-term adherence to KD remains a challenge, and its sustained impact on PD progression is unclear. In contrast, Rusch et al. [46] found that the MediDiet increased Roseburia, an anti-inflammatory bacterium, while decreasing Bilophila, thereby reducing constipation and gut inflammation. Schmit et al. [57] further emphasized the importance of dietary fiber, showing that fiber deprivation worsened PD symptoms by favoring pathogenic microbiota that promote α-synuclein aggregation. While MediDiet appears to be a more sustainable dietary intervention than KD, its direct effects on motor symptoms require further investigation.

Beyond general dietary modifications, specific metabolic factors such as branched-chain amino acids (BCAAs) and α-ketoglutarate (AKG) have been studied for their neuroprotective potential. Yan et al. [48] found that BCAA supplementation mitigated neuroinflammation, improved motor function, and preserved dopaminergic neurons in PD mouse models by reducing pro-inflammatory cytokines (IL-1β, TNF-α). However, while these preclinical findings are encouraging, their metabolic

Treatment	Mechanism	Positive outcome(s)	Limitation(s)	References
Ketogenic diet	Microbiota modulation, metabolic shifts	Improved motor function, reduced neuroinflammation	Dietary adherence challenges	[45]
Mediterranean Diet	SCFA production, gut microbiota diversity	Reduced gut inflammation, improved constipation	Unclear impact on motor symptoms	[46]
Low-protein, high-carb diet	Microbiota-metabolite-brain axis regulation	Enhanced dopamine levels, gut microbiota normalization	Long-term adherence issues	[47]
Branched-chain amino acids	Neuroinflammation regulation, energy metabolism	Improved motor function, reduced inflammation	Potential metabolic side effects	[48]
α-Ketoglutarate (AKG)	Gut microbiota modulation, DHA upregulation	Reduced α-synuclein aggregation, neuroprotection	Lacks human trials	[49]
FMT	Microbiota restoration, TLR4/NF-κB inhibition	Improved motor function, reduced inflammation	Donor selection variability, adverse effects	[35, 36]
Bacterial diet	Microbial dsRNA gene silencing	Reduced neurodegeneration	Lacks human applicability	[50]
Probiotic - *Lactobacillus salivarius* AP-32	Enhances antioxidant enzyme activity (SOD, GPx, catalase)	Protects dopaminergic neurons, improves motor function, and modulates gut microbiota	Limited human trials; long-term effects unknown	[51]
Probiotic supplement for constipation (Christensenella sp.)	Alters gut microbiota composition, improves bowel movements	Increases stool frequency, reduces constipation severity	Effects on PD progression are unknown	[52]
Probiotic combination - Microbiot® (*Bifidobacterium animalis*, *Lactobacillus rhamnosus* GG)	Reduces microglial activation and neuroinflammation	Improves motor coordination, an anti-inflammatory effect	Limited human data; unknown optimal dosage	[53]
Probiotic - *Clostridium butyricum* (Cb)	Modulates gut-brain axis via GLP-1 pathway	Improves motor deficits, enhances GLP-1 signaling, and reduces microglial activation	Mechanistic pathways are still under investigation	[54]

Treatment	Mechanism	Positive outcome(s)	Limitation(s)	References
Probiotics for immune modulation (PROB-PD trial)	Alters cytokine production and immune response	Reduces oxidative stress and inflammatory markers	Ongoing clinical trial; efficacy yet to be confirmed	[55]
Synbiotic - Enterolactis Duo (*Lacticaseibacillus paracasei DG* + inulin prebiotic)	Increases *Faecalibacterium prausnitzii*, enhances gut-brain communication	Improves mood, cognitive function, and gastrointestinal symptoms in PD patients	Short follow-up period, limited sample size	[56]
Synbiotic - Polymannuronic acid (PM) + *Lacticaseibacillus rhamnosus GG*	Enhances SCFA production, promotes GDNF and BDNF expression, and improves BBB integrity	Stronger neuroprotection than individual probiotics/prebiotics, reduces inflammation and apoptosis	Animal study; human trials needed for validation	[32]

Table 3.
Precision medicine in Parkinson's disease management: targeting gut microbiota.

effects in humans remain largely untested, and excessive BCAA intake may have unintended consequences. Similarly, Zhang et al. [49] demonstrated that AKG supplementation modified gut microbiota by increasing Lachnospiraceae, enhancing SCFA production, and reducing α-synuclein aggregation. Despite these promising findings, human clinical trials are needed to determine the therapeutic potential of AKG supplementation.

3.3.2 Fecal microbiota transplantation in Parkinson's disease management: Targeting gut microbiota

FMT has gained increasing attention as a potential strategy for restoring gut microbial balance in PD. Zhao et al. [36] demonstrated that FMT corrected gut dysbiosis in rotenone-induced PD mice, reducing neuroinflammation via modulation of the LPS-TLR4/MyD88/NF-κB signaling pathway and promoting dopaminergic neuron survival. Cheng et al. [35] extended these findings to human trials, showing that oral FMT improved PD-related autonomic symptoms and gastrointestinal disorders while enhancing microbial diversity. Despite these promising results, the long-term efficacy and optimal delivery method for FMT remain uncertain.

However, not all FMT interventions have yielded positive outcomes. Yang et al. [58] reported that transplanting gut microbiota from PD patients into healthy mice exacerbated neurodegeneration by activating the TLR4/NF-κB/NLRP3 inflammatory pathway, leading to increased α-synuclein aggregation and motor deficits. This finding highlights the importance of donor selection in FMT, as microbiota from PD patients may worsen disease pathology rather than alleviate it. Similarly, Scheperjans et al. [59] conducted a randomized controlled trial (RCT) in PD patients and found that colonoscopy-delivered FMT failed to produce significant motor improvements. Furthermore, FMT-treated patients experienced higher rates of gastrointestinal adverse effects compared to the placebo group, raising concerns about safety and the need for standardization.

3.3.3 Probiotics in Parkinson's disease management: Targeting gut microbiota

Probiotics are increasingly recognized for their potential role in managing Parkinson's disease (PD) by modulating gut microbiota and mitigating inflammation along the gut-brain axis. The gut microbiome in PD patients often exhibits dysbiosis, characterized by decreased beneficial bacterial populations and increased pro-inflammatory taxa. This imbalance is linked to oxidative stress, neuroinflammation, and gastrointestinal dysfunction, all of which contribute to PD progression.

Lactobacillus salivarius AP-32 has been shown to enhance antioxidant enzyme activity (superoxide dismutase, glutathione peroxidase, and catalase), thereby reducing oxidative stress and protecting dopaminergic neurons [51]. Additionally, probiotic treatment can restore gut microbiota balance by increasing beneficial bacteria like Christensenella, while reducing pathogenic taxa such as *Eubacterium oxidoreducens* and *Eubacterium hallii*, improving constipation symptoms in PD patients [52].

Beyond gastrointestinal improvements, probiotics exhibit immunomodulatory effects. Probiotic *Clostridium butyricum* has been found to enhance GLP-1 signaling, which reduces neuroinflammation and improves motor deficits in a mouse model of PD [54]. Similarly, the Microbiot® probiotic (*Bifidobacterium animalis* ssp. lactis Bb12 and *Lactobacillus rhamnosus* GG) has demonstrated anti-inflammatory properties by reducing microglial activation, a key contributor to neurodegeneration in PD [53].

While probiotics offer a promising, gut-targeted approach to PD management, limitations remain. The long-term effects and optimal probiotic strains are still under investigation, and clinical trials have yielded variable results due to differences in patient microbiomes, disease progression, and treatment duration. Larger, well-controlled human studies are required to fully elucidate the therapeutic potential of probiotics in PD.

3.3.4 Synbiotics in Parkinson's disease management: Targeting gut microbiota

Synbiotics, a combination of probiotics (beneficial bacteria) and prebiotics (nourishing fibers that support probiotic growth), have emerged as a promising intervention for modulating gut microbiota in Parkinson's disease (PD). Given the increasing recognition of the gut-brain axis in PD pathology, synbiotic supplementation aims to restore gut microbial balance, reduce inflammation, and enhance neuroprotection.

The Enterolactis Duo synbiotic, which contains Lacticaseibacillus paracasei DG and the prebiotic inulin, has been shown to improve non-motor symptoms in PD patients, including mood disturbances, gastrointestinal dysfunction, and cognitive performance. A clinical study found that synbiotic treatment increased beneficial gut bacteria such as *Faecalibacterium prausnitzii* and Oscillospirales while improving autonomic function and constipation [56].

In a PD mouse model, the combination of polymannuronic acid (PM) prebiotic and *Lacticaseibacillus rhamnosus* GG probiotic demonstrated stronger neuroprotective effects compared to individual treatments. PM promoted short-chain fatty acid (SCFA) production, leading to anti-inflammatory and anti-apoptotic effects, while *L. rhamnosus* GG enhanced glial cell-derived neurotrophic factor (GDNF) expression, which supports dopaminergic neuron survival. Together, these effects improved motor function and blood-brain barrier (BBB) integrity [32].

Despite these promising results, challenges remain. Variability in gut microbiota composition among PD patients, differences in individual responses to synbiotic formulations, and the lack of large-scale, long-term clinical trials limit the widespread adoption of synbiotics as a PD treatment. Future research should focus on optimizing strain selection, determining the ideal dosage, and assessing long-term neuroprotective benefits.

4. Comparison of phytotherapy, pharmacotherapy, and precision medicine in Parkinson's disease treatment

Phytotherapy, pharmacotherapy, and precision medicine each offer unique approaches to addressing gut microbiota imbalances and neurodegeneration in *Parkinson's disease (PD)*. While pharmacotherapy, particularly *levodopa-based treatments*, remains the mainstay of symptom management, emerging evidence suggests that *microbiome-modulating interventions* may enhance therapeutic efficacy and reduce side effects.

4.1 Phytotherapy vs. pharmacotherapy

A key advantage of phytotherapy over pharmacotherapy is its dual action in targeting gut dysbiosis and providing neuroprotective benefits. Unlike levodopa, which undergoes gut metabolism and is often degraded by gut bacteria, plant-based compounds such as curcumin and berberine interact with gut microbiota to either enhance dopamine synthesis or improve levodopa bioavailability. Conversely, levodopa therapy provides immediate symptom relief but contributes to gut dysbiosis and increased inflammation over time. Additionally, levodopa metabolism is affected by gut microbiota composition, which may reduce its bioavailability and effectiveness. While pharmacological interventions like levodopa-carbidopa therapy remain critical for managing PD symptoms, plant-based therapies offer a promising strategy by addressing both neuroinflammation and gut microbiota imbalances. However, phytotherapy faces challenges such as poor bioavailability, necessitating higher doses or novel delivery methods for optimal efficacy.

4.2 Pharmacotherapy vs. precision medicine

Pharmacotherapy and precision medicine differ significantly in their approach to PD treatment. Levodopa and antibiotics primarily focus on dopamine replacement and microbial modulation, but they often introduce unintended disruptions to gut microbial composition. Antibiotic interventions, such as vancomycin, have been explored for their ability to reduce neuroinflammation by modifying gut microbiota, but concerns about antibiotic resistance limit their long-term viability. In contrast, precision medicine strategies, such as probiotic supplementation, have shown potential in optimizing gut microbiota composition and improving levodopa absorption.

4.3 Precision medicine vs. phytotherapy

While phytotherapy primarily relies on plant-derived bioactive compounds to modulate gut microbiota, precision medicine employs more targeted interventions, including dietary modifications, probiotics, prebiotics, and fecal microbiota transplantation (FMT). Dietary interventions, such as the Mediterranean diet (MediDiet), promote short-chain fatty acid (SCFA)-producing bacteria, including Faecalibacterium, thereby supporting gut health and neuroprotection. FMT represents another precision medicine approach that directly restores gut microbiota composition. However, concerns about donor variability and long-term efficacy persist. Unlike FMT, plant-based interventions such as curcumin and berberine promote beneficial microbial growth while exerting anti-inflammatory effects. The challenge with phytotherapy, however, lies in poor bioavailability, whereas precision strategies such as bacteriophage therapy and FMT offer more immediate microbiota restoration.

5. Conclusion

Emerging research highlights the critical role of gut microbiota in Parkinson's disease (PD) pathology and treatment. It is important to note that phytotherapy, pharmacotherapy, and precision medicine offer distinct yet complementary approaches to modulating the gut-brain axis for improved PD management. While phytotherapy provides neuroprotective and anti-inflammatory benefits, challenges in bioavailability persist. Pharmacotherapy, particularly levodopa, remains the gold standard for symptom relief but negatively impacts gut microbiota and has limitations in long-term efficacy. Precision medicine, including probiotics, prebiotics, and FMT, enables personalized interventions that enhance treatment response while reducing side effects. Future research should prioritize integrating these strategies, exploring synergistic effects, and developing advanced delivery systems to optimize therapeutic outcomes.

Author details

Samuel Oluwadare Olalekan[1*], Patrick Godwin Okwute[2], Ifabunmi Oduyemi Osonuga[1], Adekunle Muiz Mofolorunso[2], Bukunola Oluyemisi Adegbesan[3], Esther Nkechi Ezima[3] and Gideon Olamilekan Oluwatunase[4]

1 Department of Physiology, Olabisi Onabanjo University, Ago-Iwoye, Nigeria

2 Department of Physiology, Babcock University, Ilisan-Remo, Ogun State, Nigeria

3 Department of Biochemistry, Olabisi Onabanjo University, Ago-Iwoye, Nigeria

4 Department of Anatomy, University of Medical Sciences, Ondo, Nigeria

*Address all correspondence to: samuel.olalekan@oouagoiwoye.edu.ng

References

[1] Yadav D, Kumar P. Restoration and targeting of aberrant neurotransmitters in Parkinson's disease therapeutics. Neurochemistry International. 2022;**156**:105327. Available from: https://www.sciencedirect.com/science/article/pii/S0197018622000523

[2] Feigin VL, Nichols E, Alam T, Bannick MS, Beghi E, Blake N, et al. Global, regional, and national burden of neurological disorders, 1990-2016: A systematic analysis for the Global Burden of Disease Study 2016. Lancet Neurology. 2019;**18**(5):459-480. Available from: https://www.thelancet.com/journals/laneur/article/PIIS1474-4422(18)30499-X/fulltext

[3] World Health Organization. Parkinson Disease [Internet]. Geneva, Switzerland: World Health Organization; 9 Aug 2023 [cited 2025 Feb 25]. Available from: https://www.who.int/news-room/fact-sheets/detail/parkinson-disease

[4] Leite Silva ABR, Gonçalves de Oliveira RW, Diógenes GP, de Castro Aguiar MF, Sallem CC, Lima MPP, et al. Premotor, nonmotor and motor symptoms of Parkinson's disease: A new clinical state of the art. Ageing Research Reviews. 2023;**84**:101834

[5] Kakoty V, Sarathlal KC, Dubey SK, Yang CH, Kesharwani P, Taliyan R. The gut-brain connection in the pathogenicity of Parkinson disease: Putative role of autophagy. Neuroscience Letters. 2021;**753**:135865. Available from: https://www.sciencedirect.com/science/article/pii/S0304394021002433

[6] Olalekan SO, Bakare OO, Faponle AS, Okwute PG. A review of gut microbial metabolites and therapeutic approaches in hypertension. Bulletin of the National Research Centre. 2024;**48**(1):95. DOI: 10.1186/s42269-024-01252-9

[7] Olalekan SO, Bakare OO, Osonuga IO, Faponle AS, Adegbesan BO, Ezima EN. Gut microbiota-derived metabolites: Implications for metabolic syndrome and therapeutic interventions. The Egyptian Journal of Internal Medicine. 2024;**36**(1):72. DOI: 10.1186/s43162-024-00342-4

[8] Hong JP, Shin S, Chung SH, Song CM, Shim JG, Kim Y, et al. Bacteriophages targeting *Enterococcus faecalis* enhance the therapeutic efficacy of levodopa in an MPTP-induced Parkinson's disease mouse model with *E. faecalis* gut colonization. Scientific Reports. 2024;**14**(1):26146. Available from: https://www.nature.com/articles/s41598-024-77038-w

[9] Menozzi E, Schapira AHV. The gut microbiota in Parkinson disease: Interactions with drugs and potential for therapeutic applications. CNS Drugs. 2024;**38**(5):315-331. DOI: 10.1007/s40263-024-01073-4

[10] Yao L, Yang Y, Yang X, Rezaei MJ. The interaction between nutraceuticals and gut microbiota: A novel therapeutic approach to prevent and treatment Parkinson's disease. Molecular Neurobiology. 2024;**61**(11):9078-9109. DOI: 10.1007/s12035-024-04151-2

[11] Hey G, Nair N, Klann E, Gurrala A, Safarpour D, Mai V, et al. Therapies for Parkinson's disease and the gut microbiome: Evidence for bidirectional connection. Frontiers in Aging Neuroscience. 2023;**15**:1151850. DOI: 10.3389/fnagi.2023.1151850/full

[12] Jiang L, Li J, Tang B, Guo J. Associations between gut microbiota and Parkinson disease: A bidirectional Mendelian randomization analysis. European Journal of Neurology. 2023;**30**(11):3471-3477. DOI: 10.1111/ene.15848

[13] Ramesh S, Arachchige ASPM. Depletion of dopamine in Parkinson's disease and relevant therapeutic options: A review of the literature. AIMS Neuroscience. 2023;**10**(3):200-231. Available from: https://www.ncbi.nlm.nih.gov/pmc/articles/PMC10567584/

[14] Abaza A, Jamil A, Gutlapalli SD, Ali M, Oble MJP, Sonia SN, et al. Parkinson's neuropathology puzzle: A systematic review uncovering the pathological culprits behind the neurological disease. Cureus. 2023;**15**(8):e44353

[15] Jiménez-Delgado A, Ortiz GG, Delgado-Lara DL, González-Usigli HA, González-Ortiz LJ, Cid-Hernández M, et al. Effect of melatonin administration on mitochondrial activity and oxidative stress markers in patients with Parkinson's disease. Oxidative Medicine and Cellular Longevity. 2021;**2021**:5577541

[16] Chen PC, Yu CC, Chen YS, Lu CH, Chan SH, Chou KH, et al. The potential effects of oxidative stress-related plasma abnormal protein aggregate levels on brain volume and its neuropsychiatric consequences in Parkinson's disease. Oxidative Medicine and Cellular Longevity. 2021;**2021**:3666327

[17] Chin KS, Holper S, Loveland P, Churilov L, Yassi N, Watson R. Prevalence of cerebral microbleeds in Alzheimer's disease, dementia with Lewy bodies and Parkinson's disease dementia: A systematic review and meta-analysis. Neurobiology of Aging. 2024;**134**:74-83

[18] Gandolfi M, Artusi CA, Imbalzano G, Camozzi S, Crestani M, Lopiano L, et al. Botulinum toxin for axial postural abnormalities in Parkinson's disease: A systematic review. Toxins. 2024;**16**(5):228

[19] Gautam AS, Pulivarthi CB, Singh RK. Proinflammatory IL-17 levels in serum/cerebrospinal fluid of patients with neurodegenerative diseases: A meta-analysis study. Naunyn-Schmiedeberg's Archives of Pharmacology. 2023;**396**(3):577-588

[20] Zheng Y, Li S, Yang C, Yu Z, Jiang Y, Feng T. Comparison of biospecimens for α-synuclein seed amplification assays in Parkinson's disease: A systematic review and network meta-analysis. European Journal of Neurology. 2023;**30**(12):3949-3967

[21] Proano AC, Viteri JA, Orozco EN, Calle MA, Costa SC, Reyes DV, et al. Gut microbiota and its repercussion in Parkinson's disease: A systematic review in occidental patients. Neurology International. 2023;**15**(2):750-763

[22] Vendrik KE, Chernova VO, Kuijper EJ, Terveer EM, van Hilten JJ, Contarino MF, et al. Safety and feasibility of faecal microbiota transplantation for patients with Parkinson's disease: A protocol for a self-controlled interventional donor-FMT pilot study. BMJ Open. 2023;**13**(10):e071766

[23] Adegbesan BO, Ezima EN, Hassan BO, Kehinde JO, Adewale AA, Osonuga IO, et al. Evaluation of in vitro antioxidant activities, total phenolic and elemental contents of common herbs and spices (*Moringa oleifera* leaves, *Allium sativum* (garlic) and *Momordica charantia* (ejinrin) leaves) in South-West Nigeria. Pure and Applied Chemistry. 2024;**96**(10):1339-1349. DOI: 10.1515/pac-2023-1128/html

[24] Olalekan SO, Okwute PG, Osonuga IO, Adeyanju MM. A review of botanical interventions for neuropathy and neuropathic pain. Clinical Phytoscience. 2024;**10**(1):23. DOI: 10.1186/s40816-024-00385-8

[25] Cui C, Han Y, Li H, Yu H, Zhang B, Li G. Curcumin-driven reprogramming of the gut microbiota and metabolome ameliorates motor deficits and neuroinflammation in a mouse model of Parkinson's disease. Frontiers in Cellular and Infection Microbiology. 2022;**12**:887407. DOI: 10.3389/fcimb.2022.887407/full

[26] Hall DA, Voigt RM, Cantu-Jungles TM, Hamaker B, Engen PA, Shaikh M, et al. An open label, non-randomized study assessing a prebiotic fiber intervention in a small cohort of Parkinson's disease participants. Nature Communications. 2023;**14**(1):926. Available from: https://www.nature.com/articles/s41467-023-36497-x

[27] Wang Y, Tong Q, Ma SR, Zhao ZX, Pan LB, Cong L, et al. Oral berberine improves brain dopa/dopamine levels to ameliorate Parkinson's disease by regulating gut microbiota. Signal Transduction and Targeted Therapy. 2021;**6**(1):77. Available from: https://www.nature.com/articles/s41392-020-00456-5

[28] Liu M, Xue J, Cao Y, Hao Z, Wang Y, Li J, et al. The effects of Nardosinone on levodopa intervention in the treatment of Parkinson's disease. Biomedicine & Pharmacotherapy. 2024;**174**:116448. Available from: https://www.sciencedirect.com/science/article/pii/S0753332224003329

[29] Mao Z, Hui H, Zhao X, Xu L, Qi Y, Yin L, et al. Protective effects of dioscin against Parkinson's disease via regulating bile acid metabolism through remodeling gut microbiome/GLP-1 signaling. Journal of Pharmaceutical Analysis. 2023;**13**(10):1153-1167. Available from: https://www.sciencedirect.com/science/article/pii/S2095177923001193

[30] Techaniyom P, Korsirikoon C, Rungruang T, Pakaprot N, Prombutara P, Mukda S, et al. Cold-pressed perilla seed oil: Investigating its protective influence on the gut–brain axis in mice with rotenone-induced Parkinson's disease. Food Science & Nutrition. 2024;**12**(9):6259-6283. DOI: 10.1002/fsn3.4265

[31] Cao X, Du ZR, Liu X, Wang X, Li C, Zhou SN, et al. Low and high doses of oral maslinic acid protect against Parkinson's disease via distinct gut microbiota-related mechanisms. Biomedicine & Pharmacotherapy. 2023;**165**:115100. Available from: https://www.sciencedirect.com/science/article/pii/S0753332223008910

[32] Liu X, Du ZR, Wang X, Sun XR, Zhao Q, Zhao F, et al. Polymannuronic acid prebiotic plus *Lacticaseibacillus rhamnosus* GG probiotic as a novel synbiotic promoted their separate neuroprotection against Parkinson's disease. Food Research International. 2022;**155**:111067. Available from: https://www.sciencedirect.com/science/article/pii/S0963996922001247

[33] Zhu M, Liu X, Ye Y, Yan X, Cheng Y, Zhao L, et al. Gut microbiota: A novel therapeutic target for Parkinson's disease. Frontiers in Immunology. 2022;**13**:937555. DOI: 10.3389/fimmu.2022.937555/full

[34] Cheng G, Hardy M, Hillard CJ, Feix JB, Kalyanaraman B. Mitigating gut microbial degradation of levodopa and enhancing brain dopamine: Implications in Parkinson's disease. Communications Biology. 2024;**7**(1):668. Available from: https://www.nature.com/articles/s42003-024-06330-2

[35] Cheng Y, Tan G, Zhu Q, Wang C, Ruan G, Ying S, et al. Efficacy of fecal microbiota transplantation in patients with Parkinson's disease: Clinical trial results from a randomized, placebo-controlled design. Gut Microbes. 2023;**15**(2):2284247. DOI: 10.1080/19490976.2023.2284247

[36] Zhao Z, Ning J, Bao XQ, Shang M, Ma J, Li G, et al. Fecal microbiota transplantation protects rotenone-induced Parkinson's disease mice via suppressing inflammation mediated by the lipopolysaccharide-TLR4 signaling pathway through the microbiota-gut-brain axis. Microbiome. 2021;**9**(1):226. DOI: 10.1186/s40168-021-01107-9

[37] Hong CT, Chan L, Chen KY, Lee HH, Huang LK, Yang YCS, et al. Rifaximin modifies gut microbiota and attenuates inflammation in Parkinson's disease: Preclinical and clinical studies. Cells. 2022;**11**(21):3468. Available from: https://www.mdpi.com/2073-4409/11/21/3468

[38] Cui C, Hong H, Shi Y, Zhou Y, Qiao CM, Zhao WJ, et al. Vancomycin pretreatment on MPTP-induced Parkinson's disease mice exerts neuroprotection by suppressing inflammation both in brain and gut. Journal of Neuroimmune Pharmacology. 2023;**18**(1-2):72-89. DOI: 10.1007/s11481-021-10047-y

[39] Sheng S, Li X, Zhao S, Zheng C, Zhang F. Effects of levodopa on gut bacterial antibiotic resistance in Parkinson's disease rat. Frontiers in Aging Neuroscience. 2023;**15**:1122712. DOI: 10.3389/fnagi.2023.1122712/full

[40] Zhou X, Lu J, Wei K, Wei J, Tian P, Yue M, et al. Neuroprotective effect of ceftriaxone on MPTP-induced Parkinson's disease mouse model by regulating inflammation and intestinal microbiota. Oxidative Medicine and Cellular Longevity. 2021;**2021**(1):9424582. DOI: 10.1155/2021/9424582

[41] Melis M, Vascellari S, Santoru ML, Oppo V, Fabbri M, Sarchioto M, et al. Gut microbiota and metabolome distinctive features in Parkinson disease: Focus on levodopa and levodopa-carbidopa intrajejunal gel. European Journal of Neurology. 2021;**28**(4):1198-1209. DOI: 10.1111/ene.14644

[42] Zhang Y, He X, Mo C, Liu X, Li J, Yan Z, et al. Association between microbial tyrosine decarboxylase gene and levodopa responsiveness in patients with Parkinson disease. Neurology. 2022;**99**(22):e2443-e2453. DOI: 10.1212/WNL.0000000000201204

[43] He X, Lai Y, Mo C, Zhang Y, Ai P, Xu S, et al. Association between Fecal Bile Acids and Levodopa response in patients with Parkinson's disease. Microorganisms. 2024;**12**(7):1432. Available from: https://www.mdpi.com/2076-2607/12/7/1432

[44] Baizabal-Carvallo JF, Alonso-Juarez M, Fekete R. Intestinal decontamination therapy for dyskinesia and motor fluctuations in Parkinson's disease. Frontiers in Neurology. 2021;**12**:729961. DOI: 10.3389/fneur.2021.729961/full

[45] Jiang Z, Wang X, Zhang H, Yin J, Zhao P, Yin Q, et al. Ketogenic diet protects MPTP-induced mouse model of Parkinson's disease via altering gut microbiota and metabolites. MedComm. 2023;**4**(3):e268. DOI: 10.1002/mco2.268

[46] Rusch C, Beke M, Nieves C Jr, Mai V, Stiep T, Tholanikunnel T, et al. Promotion of a Mediterranean diet alters constipation symptoms and Fecal Calprotectin in people with Parkinson's

disease: A randomized controlled trial. Nutrients. 2024;**16**(17):2946. Available from: https://www.mdpi.com/2072-6643/16/17/2946

[47] Chu C, Li T, Yu L, Li Y, Li M, Guo M, et al. A low-protein, high-carbohydrate diet exerts a neuroprotective effect on mice with 1-methyl-4-phenyl-1,2,3,6-tetrahydropyridine-induced Parkinson's disease by regulating the microbiota-metabolite–brain axis and fibroblast growth factor 21. Journal of Agricultural and Food Chemistry. 2023;**71**(23):8877-8893. DOI: 10.1021/acs.jafc.2c07606

[48] Yan Z, Yang F, Sun L, Yu J, Sun L, Si Y, et al. Role of gut microbiota-derived branched-chain amino acids in the pathogenesis of Parkinson's disease: An animal study. Brain, Behavior, and Immunity. 2022;**106**:307-321. Available from: https://www.sciencedirect.com/science/article/pii/S0889159122003841

[49] Zhang W, Ding L, Zhang M, Zheng S, Ma R, Gong J, et al. Dietary intake of α-ketoglutarate ameliorates α-synuclein pathology in mouse models of Parkinson's disease. Cellular and Molecular Life Sciences. 2023;**80**(6):155. DOI: 10.1007/s00018-023-04807-7

[50] Gaeta AL, Willicott K, Willicott CW, McKay LE, Keogh CM, Altman TJ, et al. Mechanistic impacts of bacterial diet on dopaminergic neurodegeneration in a *Caenorhabditis elegans* α-synuclein model of Parkinson's disease. IScience. 2023;**26**(6):1-27. Available from: https://www.cell.com/iscience/fulltext/S2589-0042(23)00936-7

[51] Tsao SP, Nurrahma BA, Kumar R, Wu CH, Yeh TH, Chiu CC, et al. Probiotic enhancement of antioxidant capacity and alterations of gut microbiota composition in 6-hydroxydopamin-induced Parkinson's disease rats. Antioxidants. 2021;**10**(11):1823. Available from: https://www.mdpi.com/2076-3921/10/11/1823

[52] Du Y, Li Y, Xu X, Li R, Zhang M, Cui Y, et al. Probiotics for constipation and gut microbiota in Parkinson's disease. Parkinsonism & Related Disorders. 2022;**103**:92-97. Available from: https://www.sciencedirect.com/science/article/pii/S1353802022002796

[53] Parra I, Martínez I, Vásquez-Celaya L, Gongora-Alfaro JL, Tizabi Y, Mendieta L. Neuroprotective and immunomodulatory effects of probiotics in a rat model of Parkinson's disease. Neurotoxicity Research. 2023;**41**(2):187-200. DOI: 10.1007/s12640-022-00627-y

[54] Sun J, Li H, Jin Y, Yu J, Mao S, Su KP, et al. Probiotic *Clostridium butyricum* ameliorated motor deficits in a mouse model of Parkinson's disease via gut microbiota-GLP-1 pathway. Brain, Behavior, and Immunity. 2021;**91**:703-715. Available from: https://www.sciencedirect.com/science/article/pii/S0889159120323643

[55] Martini S, Marino F, Magistrelli L, Contaldi E, Cosentino M, Comi C. The PROB-PD trial: A pilot, randomised, placebo-controlled study protocol to evaluate the feasibility and potential efficacy of probiotics in modulating peripheral immunity in subjects with Parkinson's disease. Pilot and Feasibility Studies. 2023;**9**(1):77. DOI: 10.1186/s40814-023-01306-1

[56] Andreozzi V, Cuoco S, Balestrieri M, Fierro F, Ferrara N, Erro R, et al. Synbiotic supplementation may globally improve non-motor symptoms in patients with stable Parkinson's disease: Results from an open label single-arm study. Scientific Reports.

2024;**14**(1):23095. Available from: https://www.nature.com/articles/s41598-024-74400-w

[57] Schmit KJ, Garcia P, Sciortino A, Aho VT, Rodriguez BP, Thomas MH, et al. Fiber deprivation and microbiome-borne curli shift gut bacterial populations and accelerate disease in a mouse model of Parkinson's disease. Cell Reports. 2023;**42**(9):1-28. Available from: https://www.cell.com/cell-reports/fulltext/S2211-1247(23)01082-3

[58] Yang H, Shao Y, Hu Y, Qian J, Wang P, Tian L, et al. Fecal microbiota from patients with Parkinson's disease intensifies inflammation and neurodegeneration in A53T mice. CNS Neuroscience & Therapeutics. 2024;**30**(8):e70003. DOI: 10.1111/cns.70003

[59] Scheperjans F, Levo R, Bosch B, Lääperi M, Pereira PA, Smolander OP, et al. Fecal microbiota transplantation for treatment of Parkinson disease: A randomized clinical trial. JAMA Neurology. 2024;**81**(9):925-938. Available from: https://jamanetwork.com/journals/jamaneurology/article-abstract/2821254

Chapter 4

The Role of Gut Microbiota in Cognitive Impairment

Anosh Tahir, Kiran Nooruddin, Ruquia Noor, Maryam Bakhsh and Adnan Iqbal

Abstract

This chapter focuses on the activation and modulating function of the gut-brain axis in relation to memory and other cognitive functions. The gut-brain axis is a two-way communication system between the gastrointestinal system and the brain, closely regulated by the gut microbiota. Recent studies show that gut microbiota can modulate memory through neuroinflammation, neurotransmitter interactions, and effects on synaptic plasticity. Disruption of the gut microbiota, known as gut dysbiosis, has been linked to cognitive disorders and neurodegenerative diseases. This chapter will look at the effects of the gut microbiota on memory and explain how the pathways and signaling through the microbiota affect this relationship. It will also examine therapies that are thought to regulate the composition of the human gut microbiota to improve one's memory and prevent aging. Additionally, the chapter discusses how gut-targeted therapies may be promising approaches toward enhancing cognitive health and optimal memory performance.

Keywords: gut-brain axis, gut dysbiosis, gut microbiota, memory, neurodegenerative diseases

1. Introduction

The human gut is also called the second brain and houses a complex, relatively independent network of neurons localized in the intestinal wall called the enteric nervous system (ENS) [1]. This network of neurons has a crucial role in the modulation of numerous GI functions such as motor, secretory, and circulatory blood flow [2]. Several crucial features of the ENS include its capacity for the biosynthesis and release of many types of neurotransmitters known to be used in the CNS, including serotonin, dopamine, and gamma-aminobutyric acid (GABA) [3].

Every neurotransmitter system has a distinct and crucial role to perform in retaining the normal physiological function of the nervous system. As chemical messengers between neurons, neurotransmitters act to organize various activities, including controlling mood, muscle movement, thought processes, metabolism, and more. Proper functioning of these systems is necessary for a body to be healthy, since any derailment results in several neurological and psychological issues. The cholinergic subgroup includes neurons that utilize the primary messenger acetylcholine in the transmission of nerve impulses. This neurotransmitter plays a vital role in several neurological and psychophysiological aspects, including sensory-motor functions, sleep–wake and attention-arousal systems, and cognitive behaviors related to learning and memory [4]. The glutamatergic system is made up of those neurons that contain glutamate as one of the major released neurotransmitters. This system is important in many activities of the human brain, and more so in learning and memory activities [5]. Meanwhile, the network of the GABAergic system contains neurons that release gamma-aminobutyric acid (GABA); GABA is an inhibitory transmitter that stops or prevents the continuation of some impulses to limit specific actions. It has an important function in controlling anxiety, alertness, muscle tone, learning, and epileptic activity [6].

The chemical or molecular communicators of the aminergic, monoaminergic, or biogenic amine system include dopamine, serotonin, histamine, and norepinephrine. The dopaminergic system comprises neurons that synthesize dopamine, which is an endogenous catecholamine necessary for control of behavior and physiological processes such as movement and cognition. Notably, 90% of serotonin is synthesized in the gut, which influences bowel movement and has an impact on mood and social conduct. Similarly, the dopamine secreted in the GI tract is related to the movement of the gut and, reportedly, to moods and emotions [7, 8]. A further factor that is crucial in the balance of the gut-brain axis (GBA) is the gut microbiota—a vast population of microbes that reside in the gut. These microbes are not just residents within the gut but are functioning as members of the intestinal microbiome that play an integral part in coordinating gut-brain interaction. Our microbial partners exert an effect on the brain by generating intermediate products like SCFAs and regulating neurotransmitter concentrations [9].

The gut-brain axis refers to the mechanistic pathway by which the ENS and CNS are connected and transmit signals and information between the gut and the brain, respectively. This axis is associated with neural networks such as the vagus nerve, together with endocrine and immunologic systems. Certain chemicals produced in the digestive tract can impact the mind in some ways; for example, gut serotonin affects mood and is associated with diseases like anxiety and depression. Similarly, changes in gut bacteria affect the synthesis of neuromodulators such as GABA and could influence stress and cognition [10]. Furthermore, many of these neurotransmitter interactions have ties to essential metabolic processes, underscoring how neural signaling and metabolism are interlinked within the gut.

In addition to its role in neurotransmitter regulation, the gut microbiota also plays a crucial role in metabolic processes. Literature reviews suggest that overall changes in the gut microbial community, including a decrease in microbial richness, are linked with metabolic dysregulation and obesity. Nevertheless, research findings on the relative composition of microbiota in obese subjects are somewhat contradictory. This contradiction simply cannot be overstated as it varies with other factors such as gut microbiota, diets, genes, and the surrounding environment [11]. A primary way through which the microbiota creates metabolic effects is by the production of short-chain fatty acids (SCFAs). These microbial metabolites help in maintaining energy currency through changes in some gut hormones, such as cholecystokinin (CCK), glucagon-like peptide 1 (GLP-1), peptide tyrosine-tyrosine (PYY), and leptin. These hormones govern hunger, energy management, and glucose metabolism; therefore, it is possible to assume that the gut microbiota is a main mediator of metabolism [12].

Memory and cognitive activities are central elements in human existence, since they can be supported by learning, critical thinking, and decision making. The next mental organ is the second brain located in the intestines; it has an obligatory connection with the first one—the brain; the microbiota in the gastrointestinal tract affects the brain through several links. The gut microbiota plays a crucial role in producing serotonin, dopamine, and gamma-aminobutyric acid (GABA), which are essential signaling molecules for brain function. Several studies have demonstrated that changes in gut microbiota can modulate neurotransmitter synthesis and then influence cognition and mood. That is, Lactobacillus and Bifidobacterium improve serotonin production that is crucial for regulating moods and forming memories [13, 14].

A healthy gut microbiota and a stable composition prevent excessive immune activation in the body. Chronic inflammation has been attributed to cognitive deterioration, and dysbiosis-related inflammation has been linked to neurodegenerative diseases, including Alzheimer's. This high-level overview is designed to reflect the evidence about how the gut microbiota helps in the regulation of immunities that are important in preserving and maintaining the health of the neural functions and cognitive abilities [15].

The disruption of gut microbiota, commonly called dysbiosis, or microbial dysfunction, has been associated with smaller hippocampal size and memory deficits. As previous animal experiments have also pointed out, when the gut microbiota composition has been shifted through FM, cognitive ability is affected. Human studies also back these findings by linking gut health to different cognitive disorders, including Alzheimer's and Parkinson's disease [16].

Gut microbiota is the community of microbes inhabiting the rumen and other parts of the gastrointestinal tracts that, through the gut-brain signaling, significantly affects cognition. This two-way communication net showcases neural, immune, and endocrine systems, and a new investigation underscores the importance of this axis for memory and learning as well as general cerebral function [17]. Research shows that a person's gut microbial composition correlates with the brain's efficiency, as is seen from the following studies. In middle-aged adults from the CARDIA study, the microbial community targeting β-diversity shows linear associations with better cognitive test scores across different domains, such as the Montreal Cognitive Assessment and category fluency tests. Of those, Barnesiella and Lachnospiraceae were positively correlated with better cognitive scores, and Sutterella with worse scores [18]. Based on these findings, it would be possible to state that microbial diversity plays a crucial role in protecting cognition.

Altered gut microbiota or dysbiosis has already been linked to certain cognitive decline and neurodegenerative diseases, including Alzheimer's. Chronic stress has also been known to cause alterations in gut microbiota, causing inflammation and thereby disrupting the communication channels of the body. A work using germ-free mice showed that mice that received microbiota from a stressed mouse showed reduced memory and increased anxiety-like behavior [19, 20]. Overall, these findings provide preliminary evidence that improved microbiota may help safeguard good cognitive health and counteract the deleterious effects of chronic stress.

Specifically, the gut microbiota that humans develop from infancy directly affects their brain. Sanitized animals have decreased learning abilities and poorer basal memory, pointing to the fact that microbiota is essential for proper brain functions. Consequently, the effects of probiotics, including specific strains of Lactobacillus, on memory in animal models have been demonstrated, and the research proves that the gut microbiota plays a role in early cognitive health [19, 20].

2. About the gut microbiota

The effect of the gut microbiota on human health has been recognized as an important field of study. The gut microbiota is a community of microorganisms that live in the digestive tracts of humans and animals, even insects. The human body contains the greatest number of microorganisms and the most diverse range of species in the body [21]. They are made up of thousands of microorganisms, such as bacteria, viruses, and some eukaryotes, which settle in the digestive tract soon after birth [22].

The term "human gut microbiota" can be defined as a body of microbial organisms—bacteria, fungi, and viruses—their genetic material, and waste products established in the digestive system. The gut microbiota is one of the largest and most diverse microbial communities of the human body, which is estimated to contain trillions of microbes. These microorganisms have an active role in various functions within the body, which include digestion, immune system regulation, and biosynthesis of various elements within the body [23].

The gut microbiota is also responsible for breaking down some of the intricate carbohydrates, which could not be broken down in the body independently. It assists in producing important body vitamins such as vitamin K and the B group of vitamins, besides controlling immune responses. In this way, the microbiota competes with pathogenic microbes for space and nutrients and thus helps to prevent the invasion by pathogens as part of the immune defense system [24].

Another dynamically evolving area is the fact that the gut microbiota synthesizes bioactive metabolites, in particular, SCFAs that contribute to the anti-inflammatory potential and the mucus layer of the gut. These SCFAs, and other metabolites, not only affect the local intestinal health but also physiological functions throughout the body. Further, the gut microbiota interacts with the brain through the gut-brain axis, releasing neurotransmitters, including serotonin, which influence moods, cognitive abilities, and overall mental well-being [25].

The structure of the gut microbiota is determined by the presence and density of each microbial species, which can significantly differ between people. Two major bacterial phyla that form a large part of the human gut microbiome are Firmicutes and Bacteroidetes, which have been widely investigated. It is hypothesized that the Firmicutes/Bacteroidetes (F/B) ratio might be used as a biomarker associated with specific diseases: an increased F/B ratio is proportional to obesity and metabolic disease. More specifically, Firmicutes are better designed to mine energy from a meal, while those with a high Bacteroidetes profile are generally lean [26].

Nevertheless, the F/B ratio alone does not give a complete representation of the state of the guts or the gut microbiota. Other groups of microbes that play a role in the gut include: Proteobacteria, Actinobacteria, and Verrucomicrobia. Fluctuations in the levels of these microbes have been linked to inflammatory bowel disease (IBD), irritable bowel syndrome (IBS), as well as other bowel disorders. The Microbiological profile of the microbiome may also change depending on the diseases: some microbial taxa are associated with IBD, obesity, cardiovascular disease, and even psychiatric diseases such as depression and anxiety [27].

The body needs to maintain a diverse gut microbiota because the imbalance of microbial systems contributes significantly to health problems. Quantitative data derived from microbial profiling have shown that dysbiosis is related to gastrointestinal disorders like IBS, GERD, and IBD. As for the aspect of metabolic health, dysbiosis has been linked with some disorders, such as obesity, type 2 diabetes, and cardiovascular diseases, thought to be influenced by the change in microbial fermentation profile and its subsequent impact on nutrient metabolism [28]. Also, the gut microbiome concerns directly psychological state and cognition through the production of brain hormones such as serotonin [29]. When the balance in the microbiome is disturbed, this results in psychophysical disorders that affect a person's emotional state; these include depression, anxiety, and the autistic spectrum disorder (ASD). Therefore, keeping the microbiome in a state that ensures health is imperative to digestive, metabolic, and mental health [30].

3. The gut-brain axis: A two-way communication system

"It has been thought for quite a long time that the gut and the brain are two discrete entities that do not necessarily communicate with each other but it is now known that in fact they are two integrated partners, and the ongoing 'discussion' between the two influences not only digestion but also mood, memory, and over all wellbeing" [31, 32].

The gut-brain axis (GBA) is a two-way communication system that involves the central nervous system (CNS) and the gastrointestinal tract (GIT) through neural, hormonal, and immunological signals to modulate how the body responds to stress, inflammation, and infection, while regulating the physiological functions associated with the CNS [31, 33].

This bidirectional system is between the CNS and the GIT, as shown by studies on the elements that facilitate communication between the CNS and the enteric nervous system (ENS). It comprises complex interactions through components such as the vagal nerve, the hypothalamic–pituitary–adrenal (HPA) axis, immune signaling, and the gut microbiota [34, 35].

Key components of the gut-brain axis

- *Vagus nerve*: A neuronal pathway that directly links the gastrointestinal tract and the brain.
- *Hypothalamic–pituitary–adrenal (HPA) axis*: Regulates stress responses by the transmission of endocrine signals.
- *Immune system*: Microbes and systems use cytokines and chemokines as signal molecules.
- *Gut microbiota*: Produces metabolites such as short-chain fatty acids (SCFAs) as well as neurotransmitters that have an impact on brain function [35, 36].

It could be seen that several bidirectional communication mechanisms exist between the gut microbiome, the gut, and the brain. In principle, signals from gut microbiota and specialized cells of the gut endocrine, neurocrine, and inflammation-related all reach the brain. As a result, the brain can modulate microbial communities and their activity by endocrine and neural pathways. Seven pathways of care are shown in **Figure 1** [37].

Key pathways

- *Efferent signals*: Brain→Gut—*via* neuroendocrine output, autonomic output, and HPA output.
- *Afferent signals*: Gut→Brain—by hormones, cytokines, and sensation from the gut.

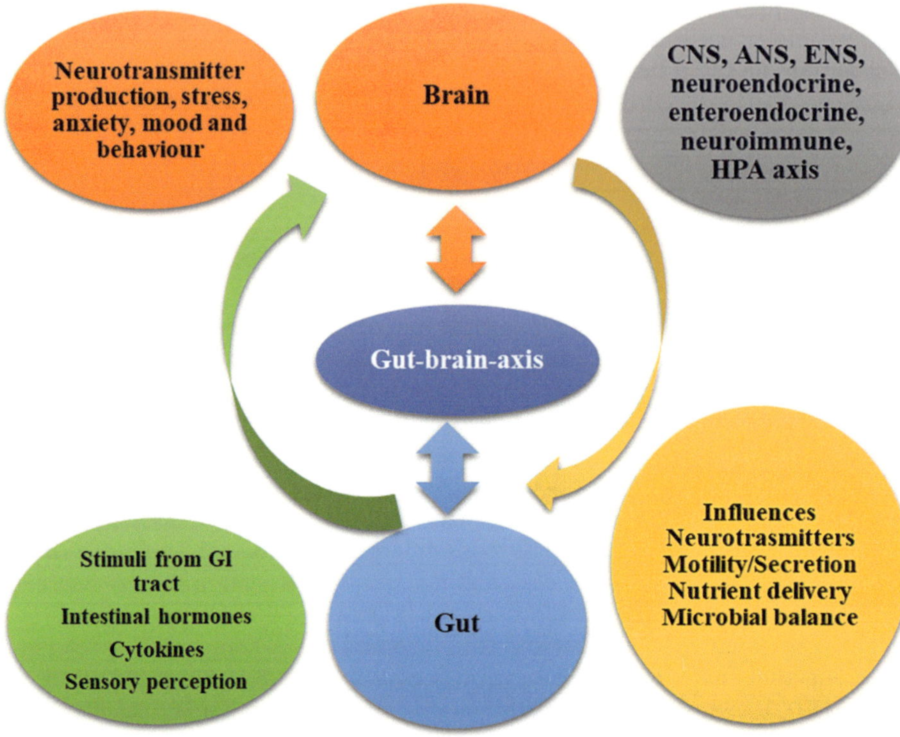

Figure 1.
The gut-brain axis represents bidirectional communication between the gut and brain, involving 'top-down' (brain-to-gut) and 'bottom-up' (gut-to-brain) pathways. These signals are mediated by systems such as the CNS, ANS, ENS, neuroendocrine, enteroendocrine, neuroimmune, and the HPA axis. In a healthy state, these pathways maintain gut and brain function. Brain-derived signals regulate gut motility, secretion, nutrient supply, and microbial balance. Conversely, gut-derived signals—such as hormones, cytokines, and sensory input—influence neurotransmitter production, stress, mood, and behavior [37].

3.1 Mechanisms of communication: Neural, hormonal, and immune pathways

The gut-to-brain axis contains two major pathways: afferents, which relay gut information, including nodose vagal and dorsal root ganglia (DRG) that convey information to the CNS through the vagus and spinal sensory nerves, respectively; herein gut-derived hormones and neurotransmitters, circulating inflammatory and immune-related signals [38].

3.1.1 Neural pathways

The neural pathways act as a basis of connecting the gut with the brain through a network of neural connections.

Enteric nervous system (ENS), also known as the second brain, is part of a connective neural tissue located in the intestinal wall. Horbacz, Pawlak, and Konturek concluded that reflex arcs within ENS regulate gut motility, secretion, and blood flow individually. These reflexes are local, but they also involve vagal and spinal afferent fibers for conveying sensory data to the CNS for synchronized digestive functions [39, 40].

The parasympathetic nervous system, for which the vagus nerve is a key part of is relied upon as the major information path between the gut and the brain. It also relays information from the gut to the brainstem, or the gut's nutrient presence or its distension. In response, the brainstem then regulates motor details falling under the gastrointestinal system, changing motility and secretion necessary [33, 41].

While vagal signaling dampens inflammation overall in the body, spinal nerves relay pain and inflammatory messages from the intestinal tract to the CNS. This pathway is especially significant in conditions where the patient is sensitive to gut stimuli, like IBS [31].

These signals that are received from the gut get to the brain stem, hypothalamus, and limbic system, therefore affecting responses to stress, moods, and even decisions made. For example, signals arising from the gut can initiate the HPA axis during stress, and there are systematic response changes [42].

In the gut-brain axis, hormonal circuits are one of the methods through which signals move from the gastrointestinal system to the CNS. Subsequent research has further refined the understanding of several hormonal integrations involving the gut-brain axis and emphasized the critical functions of the gut-born neurotransmitters and hormones in the regulation of physiological balance. Serotonin, mainly synthesized in GIT, has been studied for a long time, probably because of its multiple effects on gastrointestinal motility as well as central nervous activities. Recent studies have provided evidence that changes to the gut microbiota could have a marked effect on serotonin production and thus influence a person's moods and cognitive abilities [43, 44]. This serotonin production is controlled by the microbiota, which in turn can alter the synthesis and release of these molecules. After its synthesis, serotonin works in the brain by using the vagus nerve or by blood-borne method, regulating emotions and conduct [31].

Besides, the impact of the gut microbiota in synthesizing the inhibitory neurotransmitter gamma-aminobutyric acid (GABA) has been attracting focus. Some bacteria have been found to biosynthesize GABA, some of which affect the local gut ecosystem as well as the CNS relief [43].

In addition, interactions between ghrelin, leptin, and the CNS have been further described. Though ghrelin was initially discovered to regulate food intake, it seems to have several functions other than energy balance, including the regulation of mood and cognitive functions, which is also true for leptin, apart from its role in signaling satiety [45].

4. Gut microbiota and cognitive function

4.1 Neuroinflammation and cognitive decline

Neuroinflammation, which activates the brain's immune cells, plays a crucial role in cognitive impairment. Microglia are the resident immune cells in the brain with a critical role in promoting healthy neural responses. However, constant stimulation of these cells results in the persistent production of pro-inflammatory cytokines such as tumor necrosis factor-alpha (TNF-α), interleukin-6 (IL-6), and interleukin-1β (IL-1β). Even though these molecules are helpful in cases of acute inflammation, chronic synthesis of these molecules affects the synaptic plasticity and impairs neuron signaling and promotes neurodegeneration. This leads to inflammation, which injures neural circuits that are involved in learning and memory, and therefore contributes to the development of progressive cognitive decline [46].

One of the long-term trends that stems from the outlined neuroinflammation is the disruption of the so-called blood–brain barrier (BBB), which prevents some substances from getting into the brain. Inflammation disrupts the integrity of the BBB in chronic inflammation and allows peripheral inflammatory mediators, immune cells, and toxins to penetrate the CNS [47]. This breach also worsens inflammatory processes and increases the rate of neuronal damage. Neuroinflammation also activates the release of ROS, which leads to oxidative stress at the same time. These ROS affect neurons and synapses, leading to the aggravation of structural and functional abnormalities of the brain connectivity and selectively in the hippocampal region implicated in memory [48].

Neuroinflammation is well associated with neurodegenerative diseases, including Alzheimer's disease. These pro-inflammatory cytokines promote the formation of amyloid-beta plaques and the pathologic accumulation of hyperphosphorylated tau in Alzheimer's disease. These abnormal protein deposits exacerbate inflammation and speed up neuronal loss, which forms a dangerous feedback loop that heightens cognitive loss. Such an involvement of inflammation and neurodegenerative pathology implies that neuroinflammation should be targeted in therapeutic approaches [49].

Chronic inflammation is a prime cause since the general regulation of immunity weakens as a person grows older. Also, gut dysbiosis is an abnormal microbial ecosystem in the gut, and this affects the inflammation that flows along the gut-brain axis. Other aspects of the primary neuroinflammation include modifiable lifestyle factors, including unhealthy diet, lack of physical activity, chronic stress, and sleep deprivation, all of which contribute to additional aspects of neuroinflammation.

4.2 Neurotransmitter production and synaptic plasticity

The biogenesis of neurotransmitters and related functions in synaptic plasticity are inherent to the process of neural transmission, learning, and memory. The dorsal striatum, which forms a key part of the forebrain and constitutes a major part of what is perhaps the most intricate component of the brain involved in action control and learning, can illustrate this paradox vividly. In this area, dopaminergic, glutamatergic, and endocannabinoidic systems modulate both the afferent and efferent activities of medium spiny neurons (MSNs), the major type of neurons in the striatum. These interactions support two major forms of long-lasting synaptic plasticity, that is, long term potentiation (LTP) and long term depression (LTD) [50].

LTP at these inputs is governed by N-Methyl D-Aspartate (NMDA) receptors and may involve dopamine D1 or adenosine A2A receptors on the medium spiny neurons (MSNs). Calcium influx triggered by LTP activates calmodulin-dependent pathways (e.g., CaMKII) and calmodulin-independent mechanisms, driving synaptic modifications such as AMPA receptor insertion. These adaptations potentiate synaptic efficacy, underlying goal-directed learning and skill acquisition [51, 52].

In contrast, LTD involves retrograde endocannabinoid signaling, initiated by activation of metabotropic glutamate receptors (mGluRs) and dopamine D2 receptors [53]. This process reduces neurotransmitter release probability at presynaptic terminals, influencing the plasticity required for habit learning. This bidirectional modulation of synaptic strength (LTP and LTD) enables the striatum to support diverse learning paradigms, such as the transition from goal-directed actions to habitual behaviors as tasks are repeated and automatized [50, 54].

The striatal neuromodulation is further influenced by the interaction effect between different neurotransmitters. For example, dopamine regulates glutamate and GABA inputs in the output neurons in the striatum to regulate the excitation and inhibition required in the timing and selection of motor and other actions [55]. These endocannabinoids, acting through specific CB1 receptors, are known to play a profound role in the regulation of LTD at both the excitatory and inhibitory synapses. These mechanisms underline the basic tenets of two processes carried out by neurotransmitters, namely the direct interactions at the synaptic level and long-term adaptive forms [56].

The complex activity of neurotransmitters in the regulation of synapses and synaptic plasticity has even put them at the vanguard of cognition and behavior. Dopaminergic signaling is especially important consistently in the initial learning involving LTP and habitual learning involving LTD. The overlapping and integration of these neurotransmitter systems within the striatum allows the fine tuning of the neural circuits and thus learning and desirable behavior. These considerations also provide avenues for casting a therapeutic net for neuropsychiatric diseases characterized by abrogated plasticity, such as Parkinson's disease, which tilts the balance and undermines the graceful functions of these highly regulated circuits by virtue of aberrant dopaminergic neurotransmission [57].

5. Pathways and signaling of the gut-brain axis and memory: Bridging the gap

5.1 Immune pathways and neuroinflammation

The term neuroinflammation refers to the activation of the body's immune system response mechanisms, focusing specifically on the immune activity taking place within the CNS, and for good and bad, neuroinflammation is a process that is central to maintaining brain health. Just as acute neuroinflammation is protective through clearance of infection and repair of neural damage, chronic or uncontrolled neuroinflammation is implicated in neurodegeneration and cognitive decline. Neuroimmunology refers to the processes by which immune responses contribute to neuroinflammation, and this comprises complex interactions between resident immune cells, extravasated cytokines, and circulating inflammatory mediators.

Microglia, which are the macrophages of the central nervous system, form the fulcrum of neuroinflammation. These cells become active in response to the injury of tissue, infection, or the presence of neurotoxic substances. Once activated, the microglia assume the reactive state and begin to release presynaptic cytokines, including tumor necrosis factor-alpha (TNF-α), interleukin 6 (IL-6), and interleukin 1-beta (IL-1β). These cytokines can interfere with the normal process of synaptic transmission, hinder communication between neurons, and compromise the integrity of the BBB, allowing for the influx of additional leukocytes from the periphery and escalating neuroinflammation [58].

Toll-like receptor signaling is one of the major pathways through which an immune response is elicited in the CNS. Among those, TLR4 is engaged by LPS derived from bacterial infections, and subsequent activation of the NF-κB pathway. NF-κB migrates to the nucleus to promote the degradation of genes that manufacture further pro-inflammatory cytokines and enzymes like inducible nitric oxide synthase (iNOS) and cyclooxygenase-2 (COX-2), and sustain inflammation [59].

Oxidative stress is another feature that is evident in neuroinflammation and will be discussed next. Activated microglia release reactive oxygen species and nitric oxide, both of which have toxic effects on neurons, compromise the integrity of mitochondria, and antagonize synaptic plasticity. These reactive molecules build up in the brain and cause damage linked to neurodegenerative disorders such as Alzheimer's disease and Parkinson's disease [60].

Prostaglandins produced following the COX-2 pathway are further well-known inflammatory mediators in neuroinflammation. Related to PGE2 are neuronal excitotoxicity, synaptic dysfunction, and cognitive impairment. Blocking the COX-2 pathway can mitigate this pathway holds potential for curtailing neuroinflammation and preserving cognition [61].

Targeting immune-related pathways, which contribute to neuroinflammation, involves the use of drugs such as anti-inflammatory drugs, antioxidants, as well as those drugs that aim at reinstating the gut microbiome. For example, selective COX-2 inhibitors, proactive drugs blocking the NF-κB pathway, and probiotics normalizing systemic inflammation have been shown to precondition. Proper immune regulation and a shield against the damaging actions of constant inflammation in the brain are the goals of these approaches.

5.2 Short-chain fatty acids (SCFAs) and their role in brain and immune function

Short-chain fatty acids (SCFAs), particularly butyrate, acetate, and propionate, are metabolic byproducts synthesized by gut bacteria through the fermentation of dietary fiber and resistant starch. These metabolites play a central role in regulating communication between the gut and brain and influence nearly all aspects of neurological activity. SCFAs exert both direct and indirect effects on the central nervous system and thus are critical for overall brain function and behavior [14, 15, 19, 62].

SCFAs contribute to the functional integrity of the hippocampus—an essential region of the brain involved in memory and learning—highlighting the importance of a healthy gut microbiota for overall brain well-being. Butyrate enhances the barrier functions of the blood–brain barrier (BBB) and reduces neuroinflammation, which is essential for neuroplasticity. SCFAs are transported across the BBB through monocarboxylate transporters and help regulate brain function by modulating tight junction proteins such as claudin and occludin, thereby reducing BBB permeability and protecting the brain from inflammation [14, 15, 62, 63].

One of their key mechanisms is the regulation of neuroinflammation. SCFAs control microglial cell function, for example, butyrate reduces the levels of inflammatory cytokines such as IL-1β, TNF-α, and IL-6, and regulates the M1/M2 phenotype of microglia. These anti-inflammatory actions are partly mediated by the inhibition of histone deacetylase (HDAC), which plays a role in suppressing harmful epigenetic modifications in neurons [63, 64].

At the molecular level, SCFAs modulate neurotransmitter synthesis and neurotrophic factors. Butyrate has been shown to elevate levels of brain-derived neurotrophic factor (BDNF), a molecule essential for synaptic plasticity and memory formation. Additionally, SCFAs influence neurotransmitter pathways by acting on enzymes involved in the synthesis of serotonin, dopamine, and GABA. For instance, acetate and butyrate can upregulate proteins responsible for producing dopamine (*via* tyrosine hydroxylase) and serotonin (*via* tryptophan hydroxylase) [65].

Beyond their neurological roles, SCFAs also have significant effects on immunity. They regulate immune responses and maintain mucosal barrier integrity. These anti-inflammatory properties are especially relevant because disruptions in SCFA production have been linked to neurodegenerative diseases [66]. The immune and nervous systems interact *via* cytokines, neuropeptides, and neurotransmitters. Dysbiosis in the gut microbiota can alter cytokine profiles, contributing to the development of neuroinflammatory conditions [67, 68]. Immunocytes also synthesize neurotransmitters and neuropeptides that interact with both immune and neural receptors, supporting the concept that immune-activated signaling pathways mediated by SCFAs are integral to gut-brain communication [69].

The human gut microbiota, comprising trillions of microorganisms such as bacteria, fungi, and viruses, performs numerous essential functions, including digestion, vitamin synthesis (e.g., vitamin K and B-complex), immune regulation, and protection against pathogens by competing for nutrients and space [23, 24]. These microbes synthesize bioactive compounds like SCFAs, which affect not only local gut health but also systemic physiological and neurological processes [25].

5.3 Evidence from animal models and human studies

Animal models and human studies investigating the gut-brain axis role in cognition, mood, and memory have increased unprecedentedly. These investigations offer related views on how the microbiota, neurotransmitters, immunologic factors, and molecular communication networks are interrelated and involved in modulating the CNS and behavior.

5.3.1 Animal models

These mouse models include germ-free (GF) mice and specific-pathogen-free (SPF) mice that have played a major role in understanding the societies of gut microbiota in relation to brain health. Ablation of the microbiome results in GF mice with 50% loss of function in learning, memory, and social skills compared to SPF counterparts. For instance, animals in which GF was performed show deficits in hippocampal-dependent tasks, like spatial memory tests using the Morris water maze, and these may be attributed to low levels of BDNF in the hippocampus of GF mice. Depopulation of pathogenic microbial species and the restoration of lost friendly microbes, including Lactobacillus and Bifidobacteria, enhance learning ability and rewire the brain, proving the complexity of GI microbiota and the human brain [70, 71].

Research with both probiotics and antibiotics in animal, particularly rodent, models also points to the role of the microbiota in cognition. Probiotics have the ability to affect brain functions such as stress coping and memory, HPA axis, and cortisol levels. On the other hand, excess antibiotic use prevents healthy brain development and increases several markers of neuroinflammation, indicating how the gut microbiota affects and is affected by the central nervous system [72, 73].

5.3.2 Human studies

Cross-sectional and intervention human studies have supported those observations from animal studies and provided evidence that microbiota does play a role in cognition and mood. Research focusing on the pathophysiology of the gut-brain connection in IBS and IBD has demonstrated that changes in the composition and/or function of the gut microbiome are linked to compromise in cognitive function and increased stress reactivity [74]. Similarly, the schizophrenic patient and neuropsychiatric patients who experience depression and anxiety also show signs of gut dysbiosis, reduced serotonin synthesis, and aberrant tryptophan metabolism [75].

Studies with human subjects identified by meta-analysis of dietary interventions and of randomized controlled clinical trials with probiotics sum up well the importance of microbial communities in brain functions. For example, some food products containing *Lactobacillus rhamnosus* and *Bifidobacterium infantis* have been proven to have a positive effect on memory, decrease cortisol levels, and augmentation of the ability to regulate one's emotions. fMRI studies show alteration in the brain connectivity patterns, especially in areas involved in emotional and memory, after probiotic consumption, further extending testimonies on the efforts of microbiota in regulating cognitive functions [76, 77].

6. Link between gut dysbiosis and neurodegenerative diseases

Gut dysbiosis—the imbalance in gut microbiota—has been linked to cognitive impairment and neurodegenerative conditions like Alzheimer's and Parkinson's diseases. A number of studies published recently address the interdependence between the abnormal bacterial composition of gut microbiota and neurodegenerative diseases. The gut-brain connection ensures interaction between the gut flora and the brain, hence impacting neurological health.

6.1 Alzheimer's disease and the gut-brain axis

Cognitive impairment involving amyloid-beta plaque and neuroinflammation in AD has recently highlighted the link between the gut microbiota. Dysbiosis may contribute to AD pathogenesis through:

- *Neuroinflammation*: A disbalance in the gut microbiota induces a systemic inflammation which contributes to neuroinflammation and amyloid-beta accumulation in the brain [78, 79].

- *Microbial metabolites*: These include short-chain fatty acids, which reach the colon and are involved in the regulation of gut and brain barrier function, and are often lower in people with AD. This reduction destabilizes the neurons and their functions [80].

- *Gut barrier integrity*: In dysbiosis-associated "leaky gut," LPS translocation into the circulation activates Toll-like receptor 4 (TLR4) inflammation [81]. Currently, AD foods/herbal remedy probiotics and dietary modulations in AD are subjects of active research [82].

6.2 Parkinson's disease and gut microbiota

Parkinson's disease is a movement disorder as a result of the degeneration of the dopaminergic neurons in the substantia nigra. Evidence recommends a strong link between gut microbiota and PD:

- *Alpha-synuclein aggregation*: Dysbiosis may facilitate the accumulation of the protein known as alpha-synuclein, a critical feature of PD, in the enteric nervous system before spreading to the brain through the vagus nerve [83].

- *Gut inflammation*: A study tracing ongoing low-grade inflammation of the gut lining due to microbial dysbiosis has also highlighted a link between raised intestinal permeability and systemic inflammation and PD deterioration [84].
- *Bacterial metabolites*: The composition and the function of SCFAs related to gut microbiota changes are linked to the modulation of motor symptoms and neuroinflammation in PD patients [85]. The use of prebiotics, probiotics, and FMT as a strategy to modulate microbiota may provide positive effects on PD symptoms and possible progression.

6.3 Impact of gut microbiota disruption on memory and cognition

Imbalance in the gut microbiota mainly affects neurological processes such as memory, learning, and executive function. Mechanisms underlying these effects include:

- *Neuroinflammation*: Such inflammation results in alterations of synaptic plasticity and impairment in the hippocampal region that is important for memory [33].
- *Neurotransmitter production*: The microbiome in the gut creates hormones such as serotonin, dopamine, and GABA that have a direct impact on behavior and emotions. Impairment of the synthesis of these molecules causes dysbiosis, resulting in cognitive impairment [86].
- *HPA axis dysregulation*: Stress-induced dysbiosis disrupts the gut-brain-signaling network *via* the activation of the HPA axis, leading to cognitive impairment [87].

These studies emphasize the possibility of interventions for the modulation of the intestinal microbiota, including the use of probiotics and improved diet as means of improving cognitive function and preventing its deterioration [88].

7. Inflammation, stress, and the gut

Chronic inflammation also targets "failed" immune complexes at the blood–brain barrier, which cause significant impairments in memory and cognition, which are associated with chronic inflammation. Elevated levels of pro-inflammatory cytokines, for example, interleukin-6 (IL-6) and tumor necrosis factor-alpha (TNF-α), negatively influences neuron growth or neuroplasticity and hampers hippocampal neurogenesis, which consequently hampers cognition. Another of the studies found that elevations in IL-6 over time conferred greater risk of cognitive decline based on population-based data covering 20 years, indicating that, even at early stages, chronic inflammation has adverse effects on memory formation [89].

In addition to that, when there is always peripheral inflammation, microglia in the brain are also activated, which in turn inhibits the differentiation of neuronal stem cells and enhances apoptosis of progenitor neurons. Such a neuroinflammatory cascade disrupts the integration of new neurons developed within specific circuits and leads to memory and learning impairments [90].

As these findings show how stress and inflammation can undermine gut-brain communication, the next section will look at interventions aimed at restoring healthy gut function for better cognitive outcomes.

8. Targeting the gut for cognitive benefits

Given the significant role of gut microbiota in cognitive function, targeted interventions such as probiotics, diet modifications, and microbiome-driven therapies present exciting therapeutic possibilities. Pharmacological intervention through bacterial targeting, known as microbiota modulation, is becoming the focus for non-invasive interventions for neurological and psychiatric diseases. Sequencing techniques have made it possible to arrive at specific probiotics that can supplement one's deficiencies, likely to contribute to poor cognition or mood disorders in the gut microbiome [91]. The best-known subcategory of the probiotics is psychobiotics that act on and modulate the microbiota's bidirectional communication with the brain for improving mental health, while *Lactobacillus rhamnosus* influences the stress level and even memory stress of animals in studies conducted without using humans. Finally, postbiotics, which are products of bacterial metabolism, including SCFAs, have beneficial effects, including enhancement of the integrity of the blood–brain barrier and reduction of inflammation. These findings suggest that this approach is especially viable in neurodegenerative diseases such as Alzheimer's and Parkinson's due to a disruption of gut bacteria before the onset of symptoms. These therapies focus on the microbiome, which, if modulated, can help slow down the progression of illness through a decrease in neuroinflammation and enhanced communication between gut and brain [92].

9. Pharmacological approaches targeting gut-brain pathways

Building on the potential of gut-targeted therapies, pharmacological approaches have been developed to modulate the gut-brain axis, aiming to improve memory and cognitive processes. Pharmacological practices have been targeted at the physiological ways in which signaling interactions between the gut and the brain can promote improved memory and simpler cognitive processes, as shown in **Table 1**.

Strategy	Description	Key references
Targeting microbial metabolites	Substances that promote the synthesis of beneficial gut-derived metabolites, particularly short-chain fatty acids (SCFAs). These compounds are linked to decreased neuroinflammation and enhanced synaptic plasticity.	Frost et al. [93]
Receptor modulation	Medications acting on receptors that influence gut-brain communication, especially those affecting GI motility and cognition. Example: SSRIs target serotonin receptors in the GI tract.	Mayer et al. [33] Guzel and Mirowska-Guzel [94]
Psychobiotics development	Use of psychobiotics—a subclass of probiotics with psychopharmacological effects—to treat mood and cognitive disorders *via* gut microbiota regulation. Recent research has highlighted advances in quantifying bioactive microbial metabolites (e.g., SCFAs, GABA) that mediate gut-brain effects and supports their relevance for mental health intervention strategies.	Dinan et al. [95] Kyei-Baffour et al. [96]

Table 1.
Therapeutic strategies.

10. The microbiome's role in aging and longevity

As we explore the potential of gut-targeted therapies, it is also important to consider the role of the gut microbiota in aging and cognitive decline. The natural biological phenomenon in senescence comprises not only the gradual physical deterioration but also the various forms of cognitive constraints, for example, memory deterioration and poor control of executive functions, and a decrease in the speed of mental processes. Recent studies have established that the armies of trillions of microbes living in the gut, known as the gut microbiota, dynamically regulate these processes of aging. During aging, they found that the microbial and inflammatory properties grew more prevalent, and there was a lesser presence of various beneficial microbes such as Bifidobacteria and *Akkermansia muciniphila* [97]. These alterations modify the immune system, thereby promoting inflammation, which increases neuroinflammation, which is a major trigger to cognitive decline.

11. Conclusion

The gut-brain axis is a remarkable connection that maps the gastrointestinal tract and the brain and underlines the influence of gut microbiota on the brain and cognition. The gut microbiota can influence neurotransmitter synthesis, immune regulation, and synthesis of SCFAs and subsequently contribute to the modulation of synapse plasticity, neuroinflammation, and cognition with special reference to learning and memory phenomena.

Intestinal dysbiosis has emerged as a critical factor co-related with cognitive impairment, Alzheimer's and Parkinson's diseases, and mood disorders. The approach to gut health by management of diet, probiotics, and drugs indicates potential to boost brain health and control neuroinflammatory disorders. These therapies focus on how essential it is to preserve microbial richness and composition in the bowels for cognitive preservation and superior mental health.

Further research in this area has the ability to change the way patients with cognitive and neurodegenerative disorders are treated, as well as expand knowledge of how the brain and its biome are intricately dependent on each other.

Abbreviations

ENS	enteric nervous system
CNS	central nervous system
GABA	gamma-aminobutyric acid
SCFAs	short-chain fatty acids
HPA	hypothalamic–pituitary–adrenal (axis)
GIT	gastrointestinal tract
BBB	blood–brain barrier
LTP	long-term potentiation
LTD	long-term depression
IBS	irritable bowel syndrome
IBD	inflammatory bowel disease
CCK	cholecystokinin
GLP-1	glucagon-like peptide 1

PYY	peptide tyrosine-tyrosine
fMRI	functional magnetic resonance imaging
GERD	gastroesophageal reflux disease
DRG	dorsal root ganglia
ROS	reactive oxygen species
MSNs	medium spiny neurons
AMPA	α-amino-3-hydroxy-5-methyl-4-isoxazolepropionic acid receptor
CB1	cannabinoid receptor type 1

Author details

Anosh Tahir[1,2]*, Kiran Nooruddin[1], Ruquia Noor[1], Maryam Bakhsh[1] and Adnan Iqbal[2]

1 Dow Institute of Medical Technology, Dow University of Health Sciences, Karachi, Pakistan

2 Department of Pharmacology, Faculty of Pharmacy and Pharmaceutical Sciences, University of Karachi, Karachi, Pakistan

*Address all correspondence to: anosh.tahir@duhs.edu.pk

References

[1] Kuijer EJ, Steenbergen L. The microbiota-gut-brain axis in hippocampus-dependent learning and memory: Current state and future challenges. Neuroscience & Biobehavioral Reviews. 2023;**152**:105296

[2] Green M, Ratcliffe E. Influence of the microbiota on the development and function of the "second brain"—The enteric nervous system. In: The Gut-Brain Axis. Amsterdam, Netherlands: Elsevier; 2024. pp. 95-116

[3] Nimgampalle M, Chakravarthy H, Devanathan V. Glucose metabolism in the brain: An update. In: Recent Developments in Applied Microbiology and Biochemistry. Amsterdam, Netherlands: Elsevier; 2021. pp. 77-88

[4] Chakraborthy GS. Role of chemical agents in nervous system: A paradigm. In: Principles of Neurochemistry: Fundamentals and Applications. Cham, Switzerland; 2020. pp. 71-91

[5] Zhou Y, Danbolt NC. Glutamate as a neurotransmitter in the healthy brain. Journal of Neural Transmission. 2014;**121**:799-817

[6] Rudolph U. GABAergic system. In: Encyclopedia of Molecular Pharmacology. Berlin, Germany: Springer; 2022. pp. 679-684

[7] Juárez Olguín H, Calderón Guzmán D, Hernández García E, Barragán MG. The role of dopamine and its dysfunction as a consequence of oxidative stress. Oxidative Medicine and Cellular Longevity. 2016;**2016**(1):9730467

[8] Nimgampalle M, Chakravarthy H, Sharma S, Shree S, Bhat AR, Pradeepkiran JA, et al. Neurotransmitter systems in the etiology of major neurological disorders: Emerging insights and therapeutic implications. Ageing Research Reviews. 2023;**89**:101994

[9] Moțățăianu A, Șerban G, Andone S. The role of short-chain fatty acids in microbiota–gut–brain cross-talk with a focus on amyotrophic lateral sclerosis: A systematic review. International Journal of Molecular Sciences. 2023;**24**(20):15094

[10] Mittal R, Debs LH, Patel AP, Nguyen D, Patel K, O'Connor G, et al. Neurotransmitters: The critical modulators regulating gut–brain axis. Journal of Cellular Physiology. 2017;**232**(9):2359-2372

[11] Cryan JF, O'Riordan KJ, Sandhu K, Peterson V, Dinan TG. The gut microbiome in neurological disorders. The Lancet Neurology. 2020;**19**(2):179-194

[12] Rieder R, Wisniewski PJ, Alderman BL, Campbell SC. Microbes and mental health: A review. Brain, Behavior, and Immunity. 2017;**66**:9-17

[13] Dai W, Qiu Y, Teng Z, Li S, Huang J, Tang H, et al. Gut microbial dysbiosis and cognitive impairment in bipolar disorder: Current evidence. Frontiers in Pharmacology. 2022;**13**:893567

[14] Boem F, Ferretti G, Zipoli CS. Out of our skull, in our skin: The microbiota-gut-brain axis and the extended cognition thesis. Biology & Philosophy. 2021;**36**:1-32

[15] Sherwin E, Dinan TG, Cryan JF. Recent developments in understanding the role of the gut microbiota in brain health and disease. Annals of

the New York Academy of Sciences. 2018;**1420**(1):5-25

[16] Mayer EA, Tillisch K, Gupta A. Gut/brain axis and the microbiota. The Journal of Clinical Investigation. 2015;**125**(3):926-938

[17] Ceppa F, Mancini A, Tuohy K. Current evidence linking diet to gut microbiota and brain development and function. International Journal of Food Sciences and Nutrition. 2019;**70**(1):1-19

[18] Meyer K, Lulla A, Debroy K, Shikany JM, Yaffe K, Meirelles O, et al. Association of the gut microbiota with cognitive function in midlife. JAMA Network Open. 2022;**5**(2):e2143941-e

[19] Kraimi N, Lormant F, Calandreau L, Kempf F, Zemb O, Lemarchand J, et al. Microbiota and stress: A loop that impacts memory. Psychoneuroendocrinology. 2022;**136**:105594

[20] Thakur AK, Shakya A, Husain GM, Emerald M, Kumar V. Gut-microbiota and mental health: Current and future perspectives. Journal of Pharmacology and Clinical Toxicology. 2014;**2**(1):1016

[21] Gomaa EZ. Human gut microbiota/microbiome in health and diseases: A review. Antonie Van Leeuwenhoek. 2020;**113**(12):2019-2040

[22] Passos MCF, Moraes-Filho JP. Intestinal microbiota in digestive diseases. Arquivos de Gastroenterologia. 2017;**54**:255-262

[23] Koh A, De Vadder F, Kovatcheva-Datchary P, Bäckhed F. From dietary fiber to host physiology: Short-chain fatty acids as key bacterial metabolites. Cell. 2016;**165**(6):1332-1345

[24] Sharon G, Garg N, Debelius J, Knight R, Dorrestein PC, Mazmanian SK. Specialized metabolites from the microbiome in health and disease. Cell Metabolism. 2014;**20**(5):719-730

[25] Schroeder BO, Bäckhed F. Signals from the gut microbiota to distant organs in physiology and disease. Nature Medicine. 2016;**22**(10):1079-1089

[26] Khachatryan ZA, Ktsoyan ZA, Manukyan GP, Kelly D, Ghazaryan KA, Aminov RI. Predominant role of host genetics in controlling the composition of gut microbiota. PLoS One. 2008;**3**(8):e3064

[27] Duncan SH, Lobley G, Holtrop G, Ince J, Johnstone A, Louis P, et al. Human colonic microbiota associated with diet, obesity and weight loss. International Journal of Obesity. 2008;**32**(11):1720-1724

[28] Wang B, Yao M, Lv L, Ling Z, Li L. The human microbiota in health and disease. Engineering. 2017;**3**(1):71-82

[29] O'Mahony SM, Clarke G, Borre Y, Dinan TG, Cryan JF. Serotonin, tryptophan metabolism and the brain-gut-microbiome axis. Behavioural Brain Research. 2015;**277**:32-48

[30] Góralczyk-Bińkowska A, Szmajda-Krygier D, Kozłowska E. The microbiota–gut–brain axis in psychiatric disorders. International Journal of Molecular Sciences. 2022;**23**(19):11245

[31] Carabotti M, Scirocco A, Maselli MA, Severi C. The gut-brain axis: Interactions between enteric microbiota, central and enteric nervous systems. Annals of Gastroenterology. 2015;**28**(2):203-209

[32] Bonaz B, Bazin T, Pellissier S. The vagus nerve at the interface of the

microbiota-gut-brain axis. Frontiers in Neuroscience. 2018;**12**:1-9. DOI: 10.3389/fnins.2018.00049

[33] Mayer EA, Knight R, Mazmanian SK, Cryan JF, Tillisch K. Gut microbes and the brain: Paradigm shift in neuroscience. The Journal of Neuroscience. 2014;**34**(46):15490-15496. DOI: 10.1523/jneurosci.3299-14.2014

[34] Margolis KG, Cryan JF, Mayer EA. The microbiota-gut-brain axis: From motility to mood. Gastroenterology. 2021;**160**(5):1486-1501

[35] Ashique S, Mohanto S, Ahmed MG, Mishra N, Garg A, Chellappan DK, et al. Gut-brain axis: A cutting-edge approach to target neurological disorders and potential synbiotic application. Heliyon. 2024;**10**(13):e34092, 1-24. DOI: 10.1016/j.heliyon.2024.e34092

[36] Ahlawat S, Asha SK. Gut–organ axis: A microbial outreach and networking. Letters in Applied Microbiology. 2021;**72**(6):636-668

[37] Bajic JE, Johnston IN, Howarth GS, Hutchinson MR. From the bottom-up: Chemotherapy and gut-brain axis dysregulation. Frontiers in Behavioral Neuroscience. 2018;**12**:104

[38] Yu CD, Xu QJ, Chang RB. Vagal sensory neurons and gut-brain signaling. Current Opinion in Neurobiology. 2020;**62**:133-140. DOI: 10.1016/j.conb.2020.03.006

[39] Furness JB. The enteric nervous system and neurogastroenterology. Nature Reviews Gastroenterology & Hepatology. 2012;**9**(5):286-294. DOI: 10.1038/nrgastro.2012.32

[40] Rao M, Gershon MD. The bowel and beyond: The enteric nervous system in neurological disorders. Nature Reviews. Gastroenterology & Hepatology. 2016;**13**(9):517-528. DOI: 10.1038/nrgastro.2016.107

[41] Cryan JF, Dinan TG. Mind-altering microorganisms: The impact of the gut microbiota on brain and behaviour. Nature Reviews Neuroscience. 2012;**13**(10):701-712. DOI: 10.1038/nrn3346

[42] Gupta A, Osadchiy V, Mayer EA. Brain-gut-microbiome interactions in obesity and food addiction. Nature Reviews. Gastroenterology & Hepatology. 2020;**17**(11):655-672. DOI: 10.1038/s41575-020-0341-5

[43] Mayer EA, Nance K, Chen S. The gut–brain axis. Annual Review of Medicine. 2022;**73**(1):439-453

[44] Yano JM, Yu K, Donaldson GP, Shastri GG, Ann P, Ma L, et al. Indigenous bacteria from the gut microbiota regulate host serotonin biosynthesis. Cell. 2015;**161**(2):264-276. DOI: 10.1016/j.cell.2015.02.047

[45] Clarke GS, Page AJ, Eldeghaidy S. The gut–brain axis in appetite, satiety, food intake, and eating behavior: Insights from animal models and human studies. Pharmacology Research & Perspectives. 2024;**12**(5):e70027

[46] Fornari Laurindo L, Aparecido Dias J, Cressoni Araújo A, Torres Pomini K, Machado Galhardi C, Rucco Penteado Detregiachi C, et al. Immunological dimensions of neuroinflammation and microglial activation: Exploring innovative immunomodulatory approaches to mitigate neuroinflammatory progression. Frontiers in Immunology. 2023;**14**:1305933. DOI: 10.3389/fimmu.2023.1305933

[47] Huang X, Hussain B, Chang J. Peripheral inflammation and

blood–brain barrier disruption: Effects and mechanisms. CNS Neuroscience and Therapeutics. 2021;**27**(1):36-47

[48] Simpson DS, Oliver PL. ROS generation in microglia: Understanding oxidative stress and inflammation in neurodegenerative disease. Antioxidants (Basel, Switzerland). 2020;**9**(8):743

[49] Chen WW, Zhang X, Huang WJ. Role of neuroinflammation in neurodegenerative diseases. Molecular Medicine Reports. 2016;**13**(4):3391-3396

[50] Lovinger DM. Neurotransmitter roles in synaptic modulation, plasticity and learning in the dorsal striatum. Neuropharmacology. 2010;**58**(7):951-961. DOI: 10.1016/j.neuropharm.2010.01.008

[51] Younts TJ. Presynaptic Forms of Long-Term Plasticity in the Mammalian Hippocampus. Yeshiva University; ProQuest Dissertations & Theses Global. The University's Institutional Repository (Albert Einstein College of Medicine); 2014

[52] Annamneedi A, Gora C, Dudas A, Leray X, Bozon V, Crepieux P, et al. Towards the convergent therapeutic potential of GPCRs in autism spectrum disorders. 2022

[53] Alger BE. Retrograde signaling in the regulation of synaptic transmission: Focus on endocannabinoids. Progress in Neurobiology. 2002;**68**(4):247-286

[54] Lipton DM, Gonzales BJ, Citri A. Dorsal striatal circuits for habits, compulsions and addictions. Frontiers in Systems Neuroscience. 2019;**13**:28

[55] Do J, Kim J-I, Bakes J, Lee K, Kaang B-K. Functional roles of neurotransmitters and neuromodulators in the dorsal striatum. Learning & Memory. 2013;**20**(1):21-28

[56] Zou S, Kumar U. Cannabinoid receptors and the endocannabinoid system: Signaling and function in the central nervous system. International Journal of Molecular Sciences. 2018;**19**(3);833, 1-19. DOI: 10.3390/ijms19030833

[57] Parsons TD. Cyberpsychology and the Brain: The Interaction of Neuroscience and Affective Computing. Cambridge, United Kingdom: Cambridge University Press; 2017

[58] QQ Y, JW Z. Neuroinflammation in the central nervous system: Symphony of glial cells. Glia. 2019;**67**(6):1017-1035

[59] Zamyatina A, Heine H. Lipopolysaccharide recognition in the crossroads of TLR4 and caspase-4/11 mediated inflammatory pathways. Frontiers in Immunology. 2020;**11**:585146

[60] Fischer R, Maier O. Interrelation of oxidative stress and inflammation in neurodegenerative disease: Role of TNF. Oxidative Medicine and Cellular Longevity. 2015;**2015**(1):610813

[61] Lima IVA, Bastos LFS, Limborço-Filho M, Fiebich BL, ACP DO. Role of prostaglandins in neuroinflammatory and neurodegenerative diseases. Mediators of Inflammation. 2012;**2012**(1):946813

[62] Maiuolo J, Bulotta RM, Ruga S, Nucera S, Macrì R, Scarano F, et al. The postbiotic properties of butyrate in the modulation of the gut microbiota: The potential of its combination with polyphenols and dietary fibers. International Journal of Molecular Sciences. 2024;**25**(13):6971

[63] Dalile B, Van Oudenhove L, Vervliet B, Verbeke K. The role of short-chain fatty acids in microbiota–gut–brain communication. Nature Reviews.

Gastroenterology & Hepatology. 2019;**16**(8):461-478

[64] O'Riordan KJ, Collins MK, Moloney GM, Knox EG, Aburto MR, Fülling C, et al. Short chain fatty acids: Microbial metabolites for gut-brain axis signalling. Molecular and Cellular Endocrinology. 2022;**546**:111572

[65] Oleskin AV, Shenderov BA. Neuromodulatory effects and targets of the SCFAs and gasotransmitters produced by the human symbiotic microbiota. Microbial Ecology in Health and Disease. 2016;**27**(1):30971

[66] Zheng Y, Bonfili L, Wei T, Eleuteri AM. Understanding the gut–brain axis and its therapeutic implications for neurodegenerative disorders. Nutrients. 2023;**15**(21):4631

[67] Hou X, Artis D. Neuro-immune cell interactions in the regulation of intestinal immune homeostasis. Current Opinion in Gastroenterology. 2025;**41**(1):38-45

[68] Caldarelli M, Rio P, Marrone A, Ocarino F, Chiantore M, Candelli M, et al. Gut–brain axis: Focus on sex differences in neuroinflammation. International Journal of Molecular Sciences. 2024;**25**(10):5377

[69] Chavan SS, Ma P, Chiu IM. Neuro-immune interactions in inflammation and host defense: Implications for transplantation. American Journal of Transplantation. 2018;**18**(3):556-563. DOI: 10.1111/ajt.14515

[70] Jia P-P, Junaid M, Wen P-P, Yang Y-F, Li W-G, Yang X-G, et al. Role of germ-free animal models in understanding interactions of gut microbiota to host and environmental health: A special reference to zebrafish. Environmental Pollution. 2021;**279**:116925

[71] Tlaskalová-Hogenová H, Štěpánková R, Kozáková H, Hudcovic T, Vannucci L, Tučková L, et al. The role of gut microbiota (commensal bacteria) and the mucosal barrier in the pathogenesis of inflammatory and autoimmune diseases and cancer: Contribution of germ-free and gnotobiotic animal models of human diseases. Cellular & Molecular Immunology. 2011;**8**(2):110-120

[72] Arslanova A, Tarasova A, Alexandrova A, Novoselova V, Shaidullov I, Khusnutdinova D, et al. Protective effects of probiotics on cognitive and motor functions, anxiety level, visceral sensitivity, oxidative stress and microbiota in mice with antibiotic-induced dysbiosis. Life. 2021;**11**(8):764

[73] Fröhlich EE, Farzi A, Mayerhofer R, Reichmann F, Jačan A, Wagner B, et al. Cognitive impairment by antibiotic-induced gut dysbiosis: Analysis of gut microbiota-brain communication. Brain, Behavior, and Immunity. 2016;**56**:140-155

[74] Wang S, Zhou S, Han Z, Yu B, Xu Y, Lin Y, et al. From gut to brain: Understanding the role of microbiota in inflammatory bowel disease. Frontiers in Immunology. 2024;**15**:1384270

[75] Kim Y-K, Shin C. The microbiota-gut-brain axis in neuropsychiatric disorders: Pathophysiological mechanisms and novel treatments. Current Neuropharmacology. 2018;**16**(5):559-573

[76] Ansari F, Neshat M, Pourjafar H, Jafari SM, Samakkhah SA, Mirzakhani E. The role of probiotics and prebiotics in modulating of the gut-brain axis. Frontiers in Nutrition. 2023;**10**:1173660. DOI: 10.3389/fnut.2023.1173660

[77] Wastyk HC, Perelman D, Topf M, Fragiadakis GK, Robinson JL,

Sonnenburg JL, et al. Randomized controlled trial demonstrates response to a probiotic intervention for metabolic syndrome that may correspond to diet. Gut Microbes. 2023;**15**(1):2178794. DOI: 10.1080/19490976.2023.2178794

[78] Khatoon S, Kalam N, Rashid S, Bano G. Effects of gut microbiota on neurodegenerative diseases. Frontiers in Aging Neuroscience. 2023;**15**:1145241. DOI: 10.3389/fnagi.2023.1145241

[79] Yang J, Liang J, Hu N, He N, Liu B, Liu G, et al. The gut microbiota modulates neuroinflammation in Alzheimer's disease: Elucidating crucial factors and mechanistic underpinnings. CNS Neuroscience & Therapeutics. 2024;**30**(10):e70091. DOI: 10.1111/cns.70091

[80] Tarawneh R, Penhos E. The gut microbiome and Alzheimer's disease: Complex and bidirectional interactions. Neuroscience and Biobehavioral Reviews. 2022;**141**:104814. DOI: 10.1016/j.neubiorev.2022.104814

[81] An L, Wirth U, Koch D, Schirren M, Drefs M, Koliogiannis D, et al. The role of gut-derived lipopolysaccharides and the intestinal barrier in fatty liver diseases. Journal of Gastrointestinal Surgery. 2022;**26**(3):671-683. DOI: 10.1007/s11605-021-05188-7

[82] Vogt NM, Kerby RL, Dill-McFarland KA, Harding SJ, Merluzzi AP, Johnson SC, et al. Gut microbiome alterations in Alzheimer's disease. Scientific Reports. 2017;**7**(1):13537. DOI: 10.1038/s41598-017-13601-y

[83] Sampson TR, Debelius JW, Thron T, Janssen S, Shastri GG, Ilhan ZE, et al. Gut microbiota regulate motor deficits and neuroinflammation in a model of Parkinson's disease. Cell. 2016;**167**(6):1469-80.e12. DOI: 10.1016/j.cell.2016.11.018

[84] Zhang X, Tang B, Guo J. Parkinson's disease and gut microbiota: From clinical to mechanistic and therapeutic studies. Translational Neurodegeneration. 2023;**12**(1):59. DOI: 10.1186/s40035-023-00392-8

[85] Forero-Rodríguez J, Zimmermann J, Taubenheim J, Arias-Rodríguez N, Caicedo-Narvaez JD, Best L, et al. Changes in bacterial gut composition in Parkinson's disease and their metabolic contribution to disease development: A gut community reconstruction approach. Microorganisms. 2024;**12**(2):325

[86] Clarke G, Stilling RM, Kennedy PJ, Stanton C, Cryan JF, Dinan TG. Minireview: Gut microbiota: The neglected endocrine organ. Molecular Endocrinology. 2014;**28**(8):1221-1238. DOI: 10.1210/me.2014-1108

[87] Foster JA, Rinaman L, Cryan JF. Stress & the gut-brain axis: Regulation by the microbiome. Neurobiology of Stress. 2017;**7**:124-136. DOI: 10.1016/j.ynstr.2017.03.001

[88] Soldán M, Argalášová Ľ, Hadvinová L, Galileo B, Babjaková J. The effect of dietary types on gut microbiota composition and development of non-communicable diseases: A narrative review. Nutrients. 2024;**16**(18):3134, 1-21. DOI: 10.3390/nu16183134

[89] Wichmann MA, Cruickshanks KJ, Carlsson CM, Chappell R, Fischer ME, Klein BE, et al. Long-term systemic inflammation and cognitive impairment in a population-based cohort. Journal of the American Geriatrics Society. 2014;**62**(9):1683-1691

[90] Chesnokova V, Pechnick RN, Wawrowsky K. Chronic peripheral inflammation, hippocampal neurogenesis, and behavior. Brain, Behavior, and Immunity. 2016;**58**:1-8

[91] Nair AT, Ramachandran V, Joghee NM, Antony S, Ramalingam G. Gut microbiota dysfunction as reliable non-invasive early diagnostic biomarkers in the pathophysiology of Parkinson's disease: A critical review. Journal of Neurogastroenterology and Motility. 2018;**24**(1):30

[92] Sochocka M, Donskow-Łysoniewska K, Diniz BS, Kurpas D, Brzozowska E, Leszek J. The gut microbiome alterations and inflammation-driven pathogenesis of Alzheimer's disease—A critical review. Molecular Neurobiology. 2019;**56**:1841-1851

[93] Frost G, Sleeth ML, Sahuri-Arisoylu M, Lizarbe B, Cerdan S, Brody L, et al. The short-chain fatty acid acetate reduces appetite via a central homeostatic mechanism. Nature Communications. 2014;**5**(1):3611. DOI: 10.1038/ncomms4611

[94] Guzel T, Mirowska-Guzel D. The role of serotonin neurotransmission in gastrointestinal tract and pharmacotherapy. Molecules. 2022;**27**(5):1680. DOI: 10.3390/molecules27051680

[95] Dinan TG, Stanton C, Cryan JF. Psychobiotics: A novel class of psychotropic. Biological Psychiatry. 2013;**74**(10):720-726. DOI: 10.1016/j.biopsych.2013.05.001

[96] Kyei-Baffour VO, Vijaya AK, Burokas A, Daliri EB. Psychobiotics and the gut-brain axis: Advances in metabolite quantification and their implications for mental health. Critical Reviews in Food Science and Nutrition. 2025:1-20

[97] Shin J, Noh J-R, Choe D, Lee N, Song Y, Cho S, et al. Ageing and rejuvenation models reveal changes in key microbial communities associated with healthy ageing. Microbiome. 2021;**9**:1-19

Section 3

Traditional and Alternative Approaches to the Microbiota Problem

Chapter 5

The Role of Complementary and Alternative Therapies in the Regulation of Gut Microbiota

Mi Liu, Sihui Cao, Haihua Xie, Jia Jiang, Weiai Liu, Geshu Du, Shulin Xiong, Rong Luo and Penghui Lu

Abstract

The gut microbiota constitutes a complex ecosystem, and its composition and function are intricately linked to human health. Dysbiosis of the gut microbiota is closely associated with the onset and progression of various diseases, including gastrointestinal, immune system, metabolic, and neurological disorders. Consequently, modulating the gut microbiota to reestablish its balance has emerged as a crucial strategy for the prevention and treatment of these diseases. This chapter will delve into the roles of complementary and alternative therapies, such as acupuncture, massage, Chinese herbal medicine, dietary therapy, exercise, and psychological interventions, in modulating the gut microbiota. It will underscore the significant role of these therapies as a beneficial complement to modern medicine in the context of holistic pathology. These complementary and alternative therapies not only modulate the gut microbiota but also enhance the physiological functions of the human body and improve the body's self-healing ability through holistic regulation. These therapies offer unique advantages in clinical applications, such as fewer side effects, simple operations, and easy integration into daily life. By systematically analyzing the mechanisms of action and clinical application prospects of these therapies, this chapter aims to provide readers with a comprehensive and in-depth perspective to understand the significant value of complementary and alternative therapies in gut microbiota modulation and to offer references for future research and clinical applications.

Keywords: complementary and alternative therapies, gut microbiota, regulation, traditional medicine, holistic health

1. Introduction

The gut microbiota constitutes a complex ecosystem, teeming with trillions of microorganisms, primarily bacteria, archaea, fungi, and viruses. It is also referred to as the "second genome" of the human body. In contemporary medical research, the significance of the gut microbiota is increasingly emphasized. With the swift advancement of molecular biology technologies, particularly the widespread adoption of metagenomics and metabolomics, an increasing number of studies indicate that the composition and function of the gut microbiota are intimately connected with human health. For instance, gut microbes can break down dietary fiber, synthesize vitamins, and produce short-chain fatty acids (SCFAs), which are essential for the host's energy metabolism and gut health. Furthermore, the gut microbiota interacts with the host immune system, playing a role in the development and regulation of the immune system, and upholding the integrity of the intestinal mucosal barrier to prevent pathogen incursion.

However, the homeostasis of the gut microbiota can be affected by dietary habits, lifestyle, and psychological stress. This state of imbalance is referred to as gut dysbiosis, which is closely associated with the occurrence and development of a variety of diseases, including gastrointestinal diseases, immune system diseases, metabolic disorders, and neurological diseases. For instance, patients with inflammatory bowel disease (IBD) often exhibit significant alterations in their gut microbiota, such as a decrease in beneficial bacteria and an increase in harmful bacteria; obese individuals often show a reduction in the diversity and abundance of gut microbiota, with an increase in the number of Firmicutes and a decrease in the number of Bacteroidetes. Consequently, modulating the gut microbiota to restore its balance has become one of the important strategies for the prevention and treatment of related diseases.

Current methods for modulating the gut microbiota encompass antibiotic therapy and probiotic supplements. While these approaches can be effective to a degree, they also come with specific drawbacks. Antibiotics, frequently utilized in clinical settings to eradicate harmful bacteria, often lack specificity. They may eliminate harmful bacteria but can also harm beneficial bacteria in the process. Furthermore, prolonged antibiotic use can result in significant suppression of the microbiota and contribute to antibiotic resistance, thereby worsening the imbalance in gut microbiota and complicating treatment. Probiotic supplements aim to replenish beneficial bacteria directly, yet their effectiveness varies considerably among individuals. Their colonization and survival in the gut are limited and influenced by various factors such as gastric acid, bile acids, and the intestinal environment, and their effects are not always ideal.

Thus, in this chapter, Complementary and Alternative Medicine (CAM) has gradually garnered attention due to its safety, efficacy, and unique mechanisms of action. CAM encompasses a variety of methods and techniques employed outside the mainstream medical system to promote health and treat diseases. Within the realm of traditional Chinese medicine, complementary and alternative therapies such as Chinese herbal medicine, acupuncture, massage, dietary therapy, exercise, and psychological interventions have shown unique advantages in regulating bodily functions and enhancing the body's self-healing capabilities. These therapies influence the human body by balancing the flow of qi and blood, enhancing the functions of the internal organs, and fortifying the body's vital energy, thereby directly or indirectly affecting the composition and function of the gut microbiota. These therapies have amassed a wealth of experience through long-term clinical practice, with minimal side effects, ease of implementation, and can be seamlessly integrated into people's daily lives. Holistic pathology underscores the importance of understanding the pathogenesis and treatment strategies of diseases from a comprehensive perspective. The role of complementary and alternative therapies in modulating the gut microbiota is not only a beneficial supplement to modern medicine but also provides a new perspective for the study of holistic pathology.

This chapter will explore the influence of complementary and alternative therapies on the modulation of the gut microbiota. We will examine how traditional therapies, such as Chinese herbal medicine, acupuncture, massage, dietary therapy, exercise, and psychological interventions, can maintain and restore the balance of gut microbiota through various mechanisms, thus preventing and treating related diseases. Through a systematic analysis of these therapies, we aim to uncover their potential mechanisms in modulating gut microbiota and to investigate the prospects and challenges in their clinical application.

2. The impact of traditional Chinese medicine on the gut microbiota

Modern research has revealed that traditional Chinese medicine (TCM), as a complex system of natural remedies, offers numerous advantages. It can inhibit tumor progression, reduce therapeutic resistance, enhance immune function, and alleviate the side effects of conventional therapies. The active components of TCM modulate the composition and function of the gut microbiota through various pathways, promoting its balance and health. Thus, it serves as a reliable complementary and alternative therapy.

Studies have reported [1] the potential of Chinese herbal medicine in treating colorectal cancer, particularly through the regulation of gut microbiota and metabolites to exert anticancer effects. Research indicates that certain Chinese herbal medicines, such as Pien Tze Huang, can modulate gut microbiota and metabolites, enhance gut barrier function, and inhibit carcinogenic and pro-inflammatory pathways, thereby suppressing the occurrence of colorectal cancer. Additionally, traditional Chinese medicine formulas like Huangqin decoction and San Wu Huangqin decoction can reduce inflammation levels and improve gut microbiota imbalance by modulating gut microbiota and metabolites, thus inhibiting the development of colorectal cancer. The multi-component nature of Chinese herbal medicine provides a unique advantage in modulating gut microbiota and metabolites, offering new perspectives for the prevention and treatment of colorectal cancer.

Liu et al. [2] discussed the mechanism by which traditional Chinese medicine improves lipid metabolism disorders through the regulation of intestinal flora and their metabolites. Lipid metabolic disorders are associated with a variety of metabolic diseases, including hyperlipidemia, obesity, non-alcoholic fatty liver disease, and atherosclerosis. The imbalance of intestinal flora is considered one of the important pathogenic mechanisms of lipid metabolism disorders. In recent years, it has been found that traditional Chinese medicine, with its characteristics of multi-components and multi-targets, can significantly improve the composition and function of intestinal flora, maintain the homeostasis of intestinal flora, and regulate lipid metabolism. For instance, pachymaran can enhance intestinal barrier function and activate the peroxisome proliferator-activated receptor gamma (PPAR-γ) pathway by increasing the number of butyric acid-producing bacteria. Simiao decoction plays a role in combating non-alcoholic fatty liver disease by altering the composition of intestinal flora, particularly by increasing the proportion of *Akkermansia muciniphila* and downregulating the production of pro-inflammatory proteins. Additionally, intestinal flora can convert the effective components of traditional Chinese medicine into secondary metabolites, thus playing a therapeutic role in regulating lipid metabolism-related pathways and gene expression. The clinical application of traditional Chinese medicine and its impact on intestinal flora and lipid metabolism offer a new perspective and method for the prevention and treatment of lipid metabolism disorders (LMD).

Additionally, a study [3] isolated and purified a novel water-soluble polysaccharide, PLP1, from the root of Pueraria lobata and investigated its protective effects and mechanisms against acute alcoholic liver disease (ALD). An acute ALD model was established in mice by administering baijiu (Chinese liquor) via gavage, and the effects of PLP1 on liver injury, lipid metabolism, oxidative stress, and inflammatory responses were observed. The regulatory effects of proteolipid protein 1 (PLP1) on the gut microbiota structure and SCFA content in mice were also analyzed. The results indicated that PLP1 exhibited superior free radical scavenging ability compared to amylose and amylopectin in vitro. In vivo, PLP1 effectively protected mice from alcoholic liver injury by inhibiting oxidative stress, modulating lipid metabolism, increasing SCFA production, and maintaining gut microbiota balance. The study also found that PLP1 could regulate the gut microbiota, increasing the abundance of beneficial bacteria (such as Bacteroidetes and Lactobacillus) and reducing the abundance of harmful bacteria associated with inflammation and metabolic disorders (such as Prevotellaceae), thereby maintaining gut microbiota balance through the gut-liver axis. These findings provide a theoretical basis for the clinical application of Pueraria polysaccharides in the treatment of alcoholic liver disease and demonstrate the potential of TCM polysaccharides to exert hepatoprotective effects by modulating the gut microbiota.

A recent study [4] investigated the mechanisms through which TCM ameliorates asthma-related immune imbalances by modulating the gut microbiota. Research in recent years has indicated that the gut microbiota significantly contributes to the development of asthma. TCM regulates the gut microbiota and its metabolites, including short-chain fatty acids and lipopolysaccharides, to achieve immune regulation and suppress airway inflammation. Administered orally, TCM interacts with the gut microbiota via its active components, promoting the growth of beneficial bacteria and inhibiting the proliferation of harmful bacteria, thereby restoring the balance of the gut microbiota. Furthermore, polysaccharides, polyphenols, flavonoids, and other active components in TCM can be metabolized by the gut microbiota into substances with immune-regulatory effects, further alleviating asthma symptoms.

Numerous studies have also shown that TCM can alleviate constipation symptoms and modulate the gut microbiota and metabolites, thereby reducing inflammatory responses and protecting the intestinal mucosal barrier. For instance, Astragalus membranaceus can promote intestinal transit in loperamide-induced slow-transit constipation mice by modulating the composition of the gut microbiota and promoting the production of butyrate [5]. Another study [6] reported that Astragalus membranaceus can effectively combat constipation by altering the gut microbiota and improving the intestinal environment. Additionally, a study [7] found that Jichuan decoction, a classic and well-known traditional herbal formula composed of Angelica sinensis, Achyranthes bidentata, *Alisma plantago-aquatica*, Cistanche deserticola, Cimicifuga foetida, and Fructus Aurantii Immaturus, can alleviate chronic constipation by inhibiting the cAMP/PKA/AQPs signaling pathway, modulating inflammatory responses, and maintaining gut microbiota homeostasis. Moreover, Jichuan decoction also maintains intestinal health by reducing the apoptosis rate of enteric glial cells [8].

In recent years, the mechanisms of interaction between TCM and the gut microbiota have attracted widespread attention. Scholars believe that [9] the impact of TCM on the composition of the gut microbiota is not merely a simple summation of a few individual herbs, but rather that different combinations of herbs also play specific pharmacodynamic roles in regulating the production of gut bacterial metabolites. Other studies [10] suggest that this interaction is

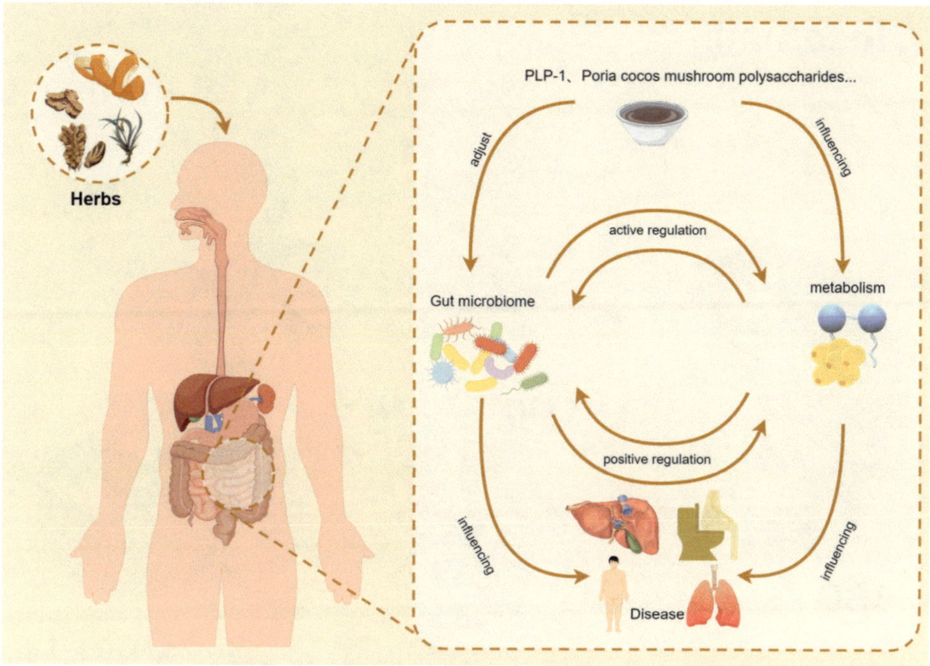

Figure 1.
The regulation of the gut microbiota by traditional Chinese medicinal herbs for disease treatment.

bidirectional: on the one hand, TCM can selectively inhibit or promote the growth of different types of gut microbes to modulate the structure and metabolic functions of the microbiota, thereby promoting human health. On the other hand, the gut microbiota can metabolize TCM, potentially enhancing its therapeutic effects or reducing its toxicity, or possibly producing toxic metabolites. However, regardless of the specific mechanism, it is clear that TCM can treat various diseases by modulating the gut microbiota (**Figure 1**).

3. The impact of acupuncture, moxibustion, and Tuina on the gut microbiota

3.1 Acupuncture

Acupuncture therapy modulates the flow of Qi and blood within the meridians by stimulating specific acupoints on the human body, thereby influencing the functions of the viscera and achieving therapeutic effects. Its characteristics are its holistic regulatory nature and non-pharmacological intervention. Compared to the commonly used drug therapies in modern medicine, acupuncture offers advantages such as being non-toxic, lacking side effects, and being easy to administer.

Research has found that acupuncture can modulate the gut microbiota, and this modulatory effect can intervene in the occurrence and development of certain specific diseases. Therefore, some of the current research progress on acupuncture's regulation of the gut microbiota in the treatment of related diseases is discussed below (**Figure 2**).

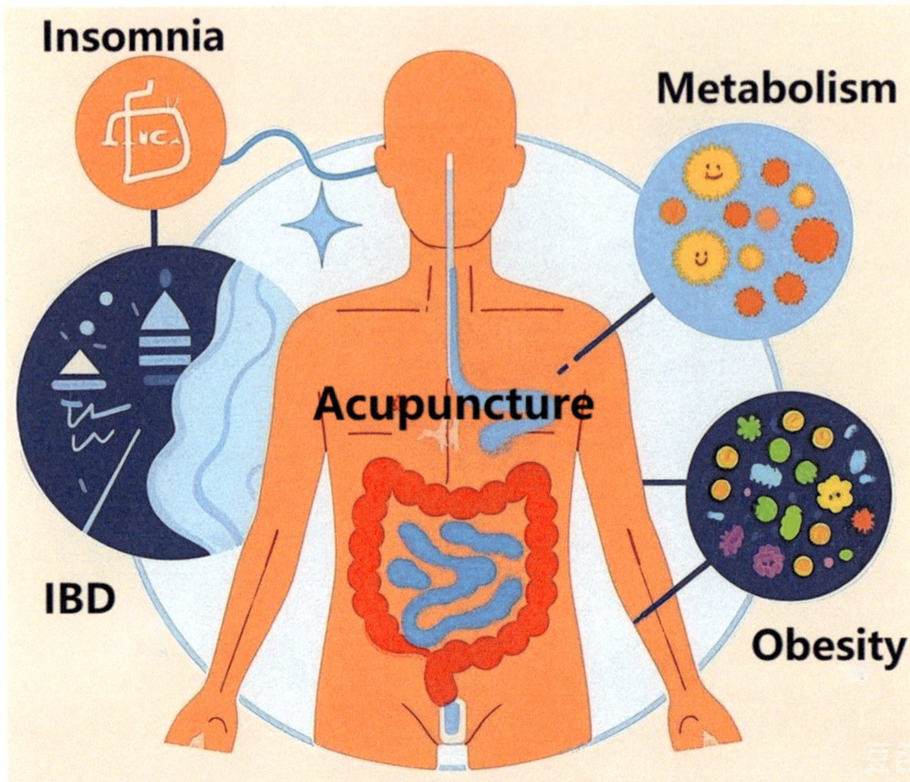

Figure 2.
Acupuncture-induced modulation of the gut Microbiota for disease treatment.

3.1.1 Acupuncture regulation of gut microbiota in insomnia model mice

Modern life is increasingly stressful, and insomnia has become the second most common mental disorder. If left untreated, chronic insomnia can harm people's physical and mental health and reduce their quality of life. Research has found [11] that acupuncture at the Baihui (DU20), Sanyinjiao (SP6), and Shenmen (HT7) acupoints for 1 minute, with an insertion depth of 1 mm, has a hypnotic and sedative effect on insomnia mice induced by p-Chlorophenylalanine (PCPA) similar to that of the drug zopiclone. This acupuncture treatment reduces the abundance of certain bacterial genera in PCPA-induced insomnia mice, such as *Faecalibacterium*, Lachnospiracea incertae sedis, *Anaerovorax*, *Oscillibacter*, Pseudoflavonifractor, and Acetatifactor, while increasing the abundance of Lactobacillus. Lachnospiracea incertae sedis, as a butyrate-producing bacterium, can trigger regulatory T cells to exert anti-inflammatory functions, thereby further regulating the immune system [12]. The abundance of Oscillibacter is significantly positively correlated with the expression of interleukin (IL)-1β and IL-6, as well as the pathological score of ulcerative colitis in mice. Lactobacillus can reduce the levels of inflammatory cytokines in mice with chronic sleep fragmentation and counteract inflammation associated with Toll-like receptors. The results suggest that acupuncture may treat insomnia by modulating the gut microbiota to regulate the host's immune response.

3.1.2 Acupuncture regulation of gut microbiota in inflammatory bowel disease model mice

Inflammatory bowel disease (IBD), encompassing ulcerative colitis (UC) and Crohn's disease (CD), is a chronic, relapsing intestinal inflammatory disorder. The prevailing theory on the pathogenesis of IBD involves the intricate interplay of genetic, environmental factors, and the host immune system, resulting in aberrant immune responses and chronic intestinal inflammation. Traditional treatments for UC include aminosalicylates, corticosteroids, antibiotics, and immunomodulators; however, their long-term use is linked to high relapse rates and severe side effects, leading to poor patient compliance. Relative to healthy individuals, IBD patients exhibit diminished gut microbiota diversity and a lower abundance of Firmicutes [13]. Studies have indicated [14] that electroacupuncture at the Guanyuan (CV4) and Zusanli (ST36) acupoints can enhance the α-diversity indices (Ace, Chao, Sobs, and Shannon) and β-diversity indices (PCA, PCoA, and NMDS) of the gut microbiota in dextran sulfate sodium (DSS)-induced UC mice, and mitigate UC symptoms. Further analysis has revealed that α-diversity indices are positively correlated with the percentage of Treg cells in CD4+ T cells and negatively correlated with the percentage of Th17 cells in CD4+ T cells. Whether electroacupuncture treatment alleviates UC symptoms via these pathways necessitates additional research. The regulation of the gut microbiota by acupuncture may emerge as a novel therapeutic approach for IBD in the future.

3.1.3 Acupuncture regulation of gut microbiota in simple obesity mice

Simple obesity, also known as primary obesity, is a prevalent condition marked by excessive fat accumulation in the body, resulting in a body mass index (BMI) that surpasses the healthy range. This can elevate the risk of developing other diseases, such as cardiovascular diseases, hypertension, arthritis, and type 2 diabetes [15]. Acupuncture is gaining recognition as a new therapeutic approach for simple obesity due to its safety, convenience, and minimal side effects. Si YC et al. [16] observed significant differences in the structure and function of the gut microbiota between healthy and obese mice. The acupuncture group's gut microbiota began to resemble that of the healthy control group. Electroacupuncture intervention led to notable improvements in basic indicators such as body weight, Lee's index, and blood lipids in patients with simple obesity. At the phylum level, the normal group was significantly enriched with Acidobacteria, Cyanobacteria, and Basidiomycota, whereas the electroacupuncture group was significantly enriched with Fusobacteria, Firmicutes, and Spirochaetes. At the species level, the normal group was significantly enriched with Bacteroides sp. CAG: 927 and Prevotella sp. CAG: 1031, while the electroacupuncture group was significantly enriched with Lachnospiraceae bacterium and Helicobacter rodentium. The alterations in bacterial diversity and metabolic genes became significant from the 7th day of electroacupuncture treatment and stabilized by the 21st day. These findings indicate that the gut microbiota could serve as a new target for electroacupuncture in treating obesity.

3.1.4 Acupuncture and the diversity of the gut microbiota

Research has indicated that acupuncture can modulate the gut microbiota, significantly enhancing its diversity and the abundance of beneficial bacteria, thereby optimizing the intestinal environment and achieving the goal of regulating the gut microbiota [17]. Xie et al. [18] discovered that acupuncture, when combined with traditional Chinese medicine, can effectively elevate the levels of Bifidobacterium and Lactobacillus in patients with ulcerative colitis, thus improving clinical symptoms. Li et al. [19] observed in a rat model of irritable bowel syndrome with diarrhea (IBS-D) that acupuncture at bilateral Tianshu, Dahuang, Zusanli, Shangjuxu, Taichong, and Baihui acupoints can increase the diversity of the gut microbiota in rats, with a rise in the abundance of Lactobacillaceae and Bifidobacteriaceae. Wang Liu-jing et al. [20] found that acupuncture at "Baihui," "Zhongwan," and "Zusanli" acupoints can decrease the gastric mucosal injury index and the extent of gastric mucosal lesions in a rat model of gastric ulcer, while also increasing the richness indices (Chao1) and diversity indices (Shannon). Wei et al. [14] treated dextran sulfate sodium-induced ulcerative colitis mice with electroacupuncture and moxibustion, and detected the gut microbiota genome using high-throughput sequencing. The findings indicated that both electroacupuncture and moxibustion improved the α-diversity and β-diversity indices of the gut microbiota.

3.2 Moxibustion

As one of the traditional Chinese medical therapies, moxibustion achieves its therapeutic effects by applying heat to acupoints along the meridians. This practice aims to warm and unblock the meridians, dispel cold, relieve pain, reduce swelling, and resolve lumps [21]. Moxibustion can also influence gastrointestinal microcirculation and regulate the gut microbiota to treat diseases [22]. There are various types of traditional Chinese moxibustion therapies. This section will discuss the research on commonly used moxibustion methods and their regulation of the gut microbiota.

Umbilical moxibustion is a technique that involves applying moxibustion to the navel, specifically the Shenque acupoint. Shenque is known for having the richest abdominal microcirculation, and the amount and rate of transdermal absorption through its skin are significantly greater than those of non-acupoint skin [23, 24]. The site of umbilical moxibustion is near the intestines, and the heat produced by its combustion can quickly penetrate into the intestines, promoting intestinal motility through warmth. Yu Ziru et al. [25] discovered that umbilical moxibustion can notably decrease the relative abundance of Escherichia-Shigella in patients with subthreshold depression (SD), while notably increasing the relative abundance of Ruminococcus and Christensenellaceae_R-7_group. It also effectively raises the relative abundance of Paraprevotella. Furthermore, they found that Christensenellaceae_R-7_group is significantly negatively correlated with the Hamilton Depression Scale (HAMD) score, whereas Escherichia-Shigella is positively correlated with the HAMD score. Umbilical moxibustion may alleviate depressive symptoms in SD patients by modulating the abundance and diversity of gut microbiota, boosting beneficial bacteria, and diminishing harmful bacteria. Jing Cai et al. [26] utilized a powder moxibustion method (consisting of Atractylodes macrocephala, Poria cocos, Atractylodes lancea, Pinellia ternata, *Syzygium aromaticum*, Borneolum syntheticum, and Zingiber officinale) in conjunction with traditional Chinese medicine health guidance to effectively enhance the phlegm-dampness constitution. This method significantly reduced the transformation score of phlegm-dampness constitution, body weight, BMI, waist circumference, and hip circumference in subjects with phlegm-dampness constitution, and notably increased the relative abundance of Eubacterium_hallii_group, with results superior to those of single health guidance. Moreover, patients with hypertension exhibit a state of gut microbiota dysbiosis, which can lead to chronic intestinal inflammation. This inflammation may increase peripheral vascular resistance and elevate blood pressure [27, 28]. Additionally, metabolites produced by the gut microbiota can influence blood pressure levels [29]. Clinical observations have indicated that [30] heat-sensitive moxibustion applied at the navel can significantly lower blood pressure in individuals with primary hypertension, particularly reducing systolic pressure. It also effectively alleviates the clinical symptoms associated with primary hypertension.

Moxibustion with medicated cakes entails applying traditional Chinese medicine to specific acupoints to harness the combined benefits of acupoints, meridians, medicinal substances, and moxibustion for therapeutic purposes [31]. Liao [32] employed moxibustion with medicated cakes to treat patients suffering from irritable bowel syndrome with diarrhea (IBS-D), leading to a significant improvement in symptoms such as frequent bowel movements, incomplete defecation, abdominal pain, and bloating. The patients' Self-Rating Anxiety Scale (SAS) scores, Self-Rating Depression Scale (SDS) scores, and Irritable Bowel Syndrome Quality of Life Scale (IBS-QOL) scores were notably reduced. This improvement may be attributed to the antibacterial properties of the medicinal substances used in this form of moxibustion, which can suppress harmful bacterial populations within the gut microbiota.

Ginger-partition moxibustion is a form of moxibustion therapy that involves placing thin slices of ginger on acupuncture points as an intermediary layer. Ginger possesses warming properties that enhance yang energy, dispel cold, and strengthen the spleen, thereby promoting intestinal blood circulation and improving intestinal function. Lin et al. et al. [33] utilized ginger-partition moxibustion to treat chronic fatigue syndrome and observed a significant increase in the abundance of the Enterobacteriaceae family within the Enterobacteriales order, the *Corynebacterium* genus in the Corynebacteriaceae family, the Erysipelotrichaceae family, and the Actinomycetales order. Additionally, the *Actinomyces* genus, *Ruminococcus* genus, and *Lactobacillus* genus exhibited a clear microbial advantage over the control group. The *Ruminococcus* genus is among the more prevalent butyrate-producing bacterial genera [34], and butyrate is crucial for maintaining the stability of the intestinal microecology, repairing the intestinal mucosal barrier, and preventing lipopolysaccharides from entering the bloodstream [35–37]. Concurrently, patients' Fatigue Scale-14 scores were notably reduced, and their fatigue symptoms were significantly alleviated, which may be attributed to the therapy's regulation of the gut microbiota structure to repair the intestinal barrier.

Heat-sensitive moxibustion is a novel form of treatment that identifies heat-sensitive acupoints using suspension moxibustion. It stimulates the arrival of moxibustion qi through specific techniques, achieving an individualized desensitization dose for each acupoint, which significantly enhances the therapeutic effect [38]. Pan et al. [39] discovered that heat-sensitive moxibustion, when combined with herbal enemas, can regulate the diversity of the gut microbiota and promote the recovery of immune function. This is particularly beneficial for the recovery of gastrointestinal function following radical gastrectomy for gastric cancer. After undergoing radical gastrectomy for gastric cancer, the beneficial bacteria Bifidobacterium and Lactobacillus were found in higher quantities in the group that received heat-sensitive moxibustion combined with herbal enemas, compared to the normal saline enema group. Simultaneously, the harmful bacteria *Escherichia coli* and Enterococcus were present in lower quantities in the same group.

In conclusion, moxibustion demonstrates significant effectiveness and safety in regulating the gut microbiota. Future research could focus on mechanistic studies and large-sample clinical trials to further explore acupoint selection, combined therapies, and the therapeutic effects of various types of moxibustion in modulating the gut microbiota. Such research would hold greater guiding and practical significance for both scientific inquiry and clinical practice (**Figure 3**).

3.3 Tuina

Tuina is a traditional manual therapy rooted in traditional Chinese medicine theories, such as meridian theory and the concepts of Qi and blood. It also incorporates modern medical knowledge. Tuina influences specific parts of the human body surface, including acupoints, meridians, muscles, and joints, through manual operations like pressing, rubbing, pushing, kneading, and pinching. The aim is to regulate the flow of Qi and blood, unblock meridians, balance Yin and Yang, relieve pain, and improve function. Thus, it achieves the goals of disease prevention and treatment, health promotion, and rehabilitation.

The origins of Tuina date back to ancient times when people instinctively relieved discomfort by pressing and kneading painful areas of the body, marking the primitive form of Tuina. In modern times, Tuina has been integrated with modern medical knowledge, becoming more scientific and standardized. It has been included in the educational curriculum of traditional Chinese medicine colleges and is widely applied in clinical practice. Additionally, Tuina has gained international recognition and has become an integral part of the CAM field. It is applicable not only to diseases of the digestive system, musculoskeletal system, nervous system, and pediatrics, as well as rehabilitation treatment, but is also extensively used in health care and wellness.

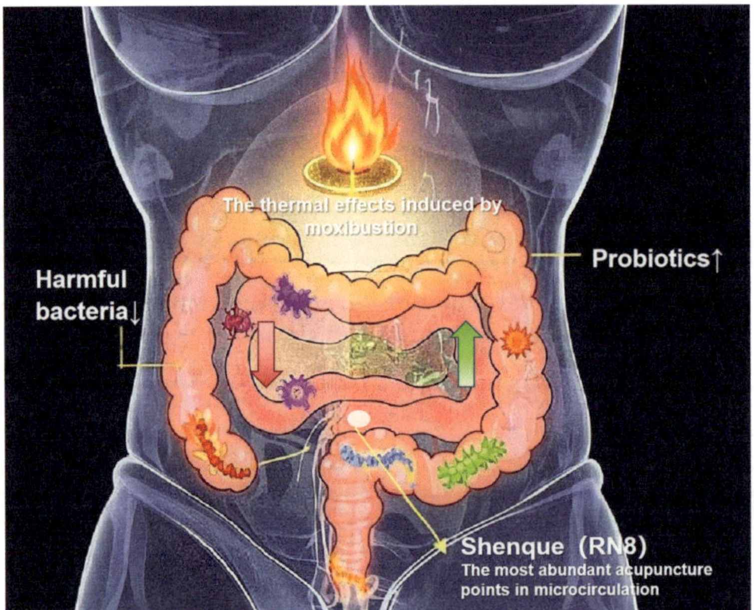

Figure 3.
Moxibustion-induced modulation of the gut microbiota for disease treatment.

Tuina has shown potential in modulating the gut microbiota in various diseases. Preclinical studies have revealed that abdominal massage can increase the number of beneficial bacteria (such as the Bifidobacterium genus) while reducing the levels of harmful bacteria (such as the Ruminococcus_torques_group and Clostridium_innocuum_group), significantly improving symptoms of functional diarrhea, restoring intestinal barrier function, and reversing gut microbiota imbalance [40]. Tuina therapy can also enhance vagal activity and promote the secretion of appetite-regulating factors, thereby enhancing gastrointestinal motility and improving digestive function, while modulating gut microbiota activity, effectively treating anorexia [41].

Furthermore, preclinical studies have indicated that Tuina significantly enhances the Shannon index of the gut microbiota in rats. Abdominal Tuina effectively promotes gastrointestinal motility, enriches the diversity of gut microbiota, and boosts gut microbiota activity, thereby alleviating symptoms of slow-transit constipation (STC) induced by colonic motility inhibitors in rats and facilitating disease resolution [42]. Concurrently, clinical studies have demonstrated that acupoint Tuina therapy can modulate the gut microbial community structure in patients with chronic functional constipation. The findings suggest that Tuina therapy increases beneficial bacterial genera, such as Pseudobutyrivibrio and Ruminiclostridium, while decreasing the harmful bacterial genus Fusicatenibacter, thus alleviating constipation symptoms. The potential mechanisms may involve the modulation of metabolic pathways and alterations in immune responses [43].

Recent research has indicated that Tuina therapy holds significant potential and clinical efficacy in treating metabolic diseases. For instance, in the context of Type 2 Diabetes Mellitus (T2DM), abdominal vibration Tuina has been shown to modulate the gut microbiota structure of T2DM rats. This is primarily evidenced by an increase in the relative abundance of Bacteroidetes; a reduction in the Firmicutes to Bacteroidetes ratio; an elevation in the relative abundance of beneficial bacteria such as Bifidobacterium and Akkermansia; a decrease in the relative abundance of opportunistic pathogens including Proteobacteria and its related classes and orders, Corynebacterium; and a reduction in the relative abundance of bacteria that negatively impact health, such as Blautia, Bacteroides wadsworthia, Ruminococcus, and Desulfovibrio. Abdominal vibration Tuina promotes the production of SCFAs, enhances intestinal barrier function, positively regulates the gut microbiota, and modulates intestinal homeostasis by altering the gut microbiota structure [44]. Clinical research data demonstrate that abdominal massage notably increases the abundance of beneficial bacteria (such as *Bifidobacterium* and *Lactobacillus*) while decreasing the abundance of harmful bacteria (such as Enterobacteriaceae and Enterococcus). This intervention corrects the gut microbiota disorder in T2DM patients to a significant degree. It exhibits substantial benefits in improving blood glucose control and lipid metabolism, including a reduction in total cholesterol, among T2DM patients [45].

Preclinical studies have demonstrated that Tuina therapy can significantly enhance social deficits, alleviate anxiety symptoms, and diminish stereotyped behaviors in autism spectrum disorder (ASD) model rats by altering the gut microbial community structure. This includes increasing the population of beneficial bacteria such as Blautia and decreasing the presence of harmful bacteria like Lactobacillus, as well as enhancing the metabolic functions of neurotransmitters [46]. Clinical studies have further confirmed that abdominal vibration Tuina can lower the Firmicutes/Bacteroidetes ratio in the gut microbiota of children with autism, elevate the proportion of probiotics including Bifidobacterium, and reduce the proportion of Bacteroides and Ruminococcus. These changes positively impact gastrointestinal function and behavioral performance in children with ASD, potentially linked to improvements in inflammatory response levels in these patients [47].

In summary, as a form of complementary and alternative therapy, Tuina has exhibited promising effects and potential applications in modulating the gut microbiota and alleviating various diseases.

4. The impact of dietary intervention on the gut microbiota

Diet is a decisive factor among the many that influence gut microbiota. A classic statement published in a prestigious journal once declared, "You are what you eat," succinctly capturing the close relationship between diet and gut microbiota.

The types and quantities of gut microbiota differ among individuals, but they can typically be categorized into three groups: beneficial bacteria, harmful bacteria, and opportunistic pathogens. Different dietary habits can markedly alter the composition and function of the gut microbiota, thereby impacting human health. For instance, a high-fiber diet can boost the number and diversity of beneficial bacteria, whereas a diet high in fat and protein may increase the prevalence of harmful bacteria. Dietary interventions on gut microbiota offer several advantages, including rapid action, affordability, absence of side effects, and sustainability, making them an effective adjunct therapy and a means for long-term regulation. Some scholars have discovered that [48] resveratrol can increase the ratio of Bacteroidetes to Firmicutes in the gut of mice, inhibit the growth of the Prevotella genus, and enhance the abundance of Bacteroides, Lactobacillus, Bifidobacterium, and Akkermansia. By altering the gut microbiota, it suppresses the production of trimethylamine, thereby reducing the levels of trimethylamine N-oxide (TMAO) and lowering the risk of cardiovascular disease. Diet can significantly alter the composition of the gut microbiota within 24 hours, and the changes in the gut microbiota caused by long-term dietary intervention are more stable and difficult to reverse, and may even be irreversible. Studies have shown that [49] feeding mice with low-dietary fiber food leads to a decrease in gut microbiota diversity, but the changes in the gut microbiota can be effectively reversed after being fed with food rich in dietary fiber.

4.1 Applications of dietary intervention in different diseases

4.1.1 Cardiovascular diseases

Hypertension is a high-risk factor for cardiovascular diseases. There is an interaction between dietary salt intake, gut microbiota composition, and blood pressure. A high-fiber diet can improve gut microbiota dysbiosis, increase the levels of acetogenic and Bacteroidetes, and enhance the local barrier function of the intestinal mucosa, thereby lowering blood pressure and reducing the risk of cardiovascular diseases [50].

The prevention and treatment of cardiovascular diseases can begin with the modulation of gut microbiota, with diet being the fundamental means. The first step involves improving one's daily diet by reducing the intake of animal products, particularly red meat, and incorporating extra virgin olive oil, balsamic vinegar, and grapeseed oil while adhering to a Mediterranean diet. This diet, primarily consisting of fish, legumes, vegetables, fruits, nuts, and olive oil, with moderate red wine consumption, is believed to prevent cardiovascular diseases and decrease the mortality rate associated with them [51]. In the case of coronary heart disease, transitioning from a diet high in red meat to one rich in white meat or non-meat protein can significantly reduce TMAO levels within a few weeks, thereby inhibiting its role in promoting coronary atherosclerotic heart disease.

4.1.2 Tumor

Dysbiosis of the gut microbiota can lead to the disruption of the intestinal mucosa and an increase in inflammatory responses, thereby increasing the risk of cancer. Dietary intervention can help patients improve their nutritional status, enhance their quality of life, and by altering the levels of metabolites within the body, affect the metabolic processes of tumor cells, thereby improving the effectiveness of cancer treatment. The ketogenic diet can enrich specific gut microbiota, such as *Akkermansia muciniphila* and *Bifidobacterium adolescentis*, which produce 3-hydroxybutyrate. 3-hydroxybutyrate can increase the proliferation of cytotoxic T cells induced by immune checkpoint inhibitors and inhibit the expression of programmed death ligand 1, thereby maintaining T cell activation and exerting anti-tumor effects, enhancing the efficacy of immune checkpoint inhibitors [52]. Vitamin D, by acting on the vitamin D receptor of intestinal epithelial cells, promotes the enrichment of the gut bacterial genus *Bacteroides fragilis*, stimulates cytotoxic T cells to produce interferon-γ, and enhances the immune response, thereby increasing the body's immunity to cancer, enhancing the effects of immune checkpoint inhibitors, and inhibiting tumor growth [53].

4.1.3 Neurological and mental disorders

There is a certain correlation between gut microbiota and mental disorders; probiotics are beneficial to mental health, whereas pathogenic bacteria may be implicated in the development of mental disorders. Dysbiosis of the gut microbiota can lead to gastrointestinal abnormalities, such as diarrhea and constipation, and may also cause psychological symptoms like anxiety and depression. The ketogenic diet can significantly improve symptoms of autism spectrum disorder and is beneficial to patients with autism spectrum disorder [54].

Healthy dietary patterns, including the Mediterranean diet, high-fiber diet, and ketogenic diet, can improve the composition of the gut microbiota and positively affect behavior and cognition through glucocerebrosidase, which helps to alleviate symptoms of depression [55]. Diets rich in ω-3 polyunsaturated fatty acids and vitamin A have a preventive effect on stress-induced cognitive behavior and gut microbiota dysbiosis [56].

4.1.4 Metabolic diseases

The gut microbiota can influence the development of metabolic diseases such as energy balance, obesity, and diabetes in the host through its metabolic activities. The gut microbiota mainly affects the host's metabolic processes by producing metabolites such as SCFAs and lactic acid, which can influence the host's carbohydrate and lipid metabolism. A low-protein diet and low-carbohydrate intake are conducive to the formation of a healthy gut microbiota and can improve the glucose tolerance levels of patients with type 2 diabetes [57]. Supplementing the gut of celiac disease patients with probiotics can regulate the composition and function of the gut microbiota

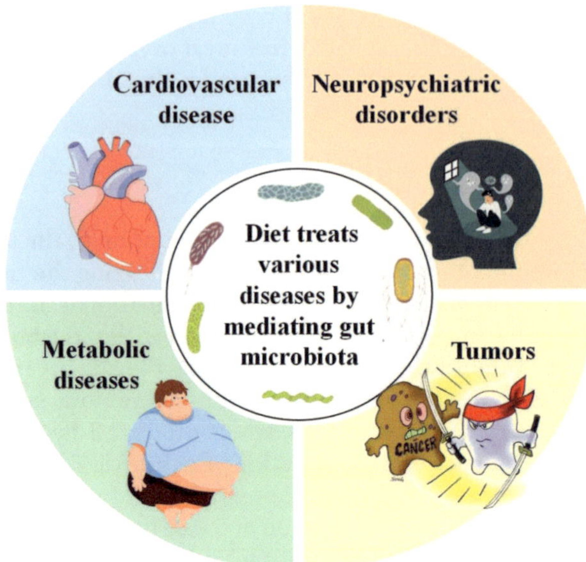

Figure 4.
Diet treats various diseases by mediating gutmierobiota.

through the bacteria themselves or their metabolites, reduce the immunogenicity of gliadin peptides, maintain the integrity of the intestinal barrier, and repair mucosal immune disorders and other pathological conditions [58].

In summary, diet and gut microbiota are closely related. On the one hand, unhealthy dietary patterns can alter the composition of the microbiota, thereby influencing the onset of diseases. On the other hand, dietary intervention is also an important method for preventing and treating diseases. Targeted modulation of the gut microbiota through daily dietary supplements is a trend in the development of the healthy food industry for the future (**Figure 4**).

5. The impact of exercise on the gut microbiota

In TCM philosophy, the equilibrium of the gut microbiota is intimately connected to the functions of the spleen and stomach, the circulation of Qi and blood, the harmony between Yin and Yang, and the coordinated functioning of the five viscera and six bowels. TCM posits that gut health encompasses not only the material process of digestion and absorption but also involves the transformation of Qi and blood, the nourishment of bodily fluids, and the regulation of Yin and Yang.

Thus, through the practice of Traditional Chinese Medicine (TCM) exercises, the balance and health of the gut microbiota can be promoted in the following ways: Firstly, the spleen and stomach are considered the "foundation of acquired constitution" in TCM, being the source of the transformation of Qi and blood. The spleen governs transportation and transformation, while the stomach is responsible for receiving and accommodating food. Through TCM exercises, such as Qigong and Tai Chi, the functions of the spleen and stomach can be regulated, and the health of the digestive system can be improved. For instance, the stretching and relaxation inherent in Tai Chi movements assist in unblocking the Qi mechanisms of the spleen and stomach, enhancing their transportation and transformation functions, thereby providing a more favorable digestive environment for the gut and contributing to the balance of the gut microbiota. Secondly, the abundance of Qi and blood is fundamental to gut health. The movement of Qi facilitates the flow of blood, and adequate Qi and blood can nourish the five viscera, sustaining the normal functioning of the gut. Through practices such as Qigong and Tai Chi, the circulation of Qi and blood can be regulated, the meridians can be unblocked, and the occurrence of Qi and blood stagnation can be prevented, thus maintaining the normal function of the gut and the balance of healthy microbiota. Thirdly, TCM posits that an imbalance between Yin and Yang within the human body can result in various diseases. The health of the gastrointestinal system also necessitates the maintenance of Yin-Yang equilibrium. Engaging in practices such as Tai Chi and Qigong not only harmonizes the body's Yin and Yang but also strengthens the Yang Qi of the spleen and stomach, facilitating the proper circulation of Yang Qi throughout the body. Consequently, this sustains the stability of the gastrointestinal tract's internal environment. For instance, a constitution with insufficient Yang may cause sluggish intestinal motility. By enhancing Yang Qi through TCM exercises, gut function can be improved and the proliferation of beneficial bacteria can be promoted.

5.1 Baduanjin

Baduanjin, a form of fitness Qigong, was created by the General Administration of Sport of China. It features simple movements and graceful postures, with an appropriate level of exercise intensity, making it a typical aerobic exercise. From a microbiological perspective, studies have been conducted on the gut microbiota of elderly individuals who regularly practice Baduanjin. The results indicate that long-term practice substantially boosts the population of beneficial bacteria, including Bifidobacterium and Lactobacillus, within the gut of the elderly. This indicates that the Baduanjin exercise promotes the proliferation of beneficial bacteria, thereby enhancing the stability and physiological functions of the gut. It also boosts resistance to harmful bacteria and improves the intestinal micro-ecological environment in the elderly, contributing to better health.

5.2 Wuqinxi

Current research has found that [59] exercise can positively regulate the structure of the gut microbiota. For instance, the gut microbiota diversity of Irish rugby players has significantly changed at multiple taxonomic levels, and the gut microbiota of individuals who exercise regularly is more diverse, with an increase in the abundance of Firmicutes. Developing good exercise habits helps to increase the number of beneficial bacteria in the body while reducing harmful bacteria. In China, exercises such as Baduanjin and Tai Chi can improve the gut microbiota condition of the elderly population.

Building on the aforementioned research findings, significant progress has been made in the study of Wuqinxi: researchers have explored the impact of Wuqinxi on middle-aged men with metabolic syndrome (MS) and healthy individuals at the genus level. The results indicate that Wuqinxi intervention can effectively optimize gut microbiota, exhibiting a clear time-dose effect. As the intervention period extends, indicators of beneficial bacteria, such as Bifidobacterium and Lactobacillus, and their related metabolites, show an upward trend, while harmful bacteria, including

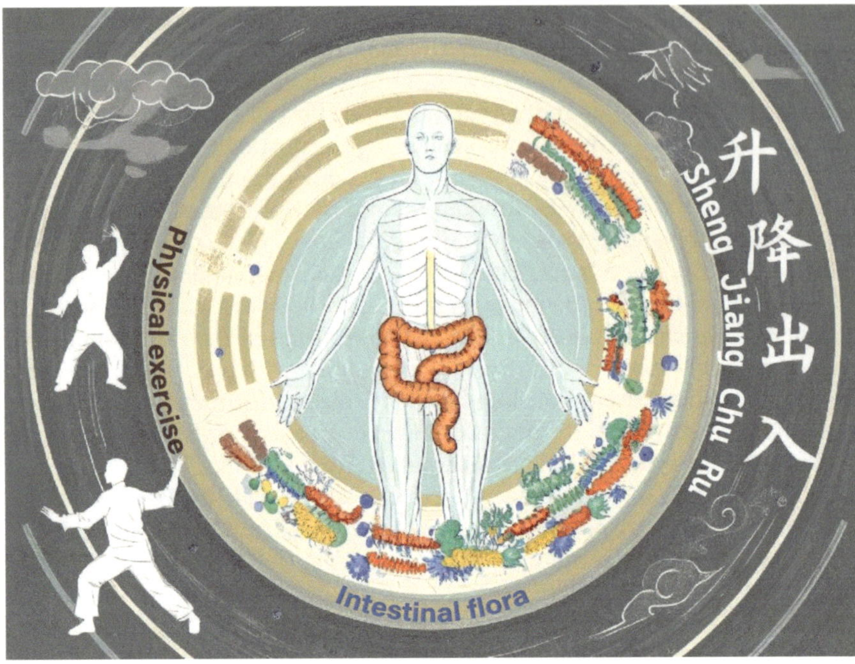

Figure 5.
Exercise-induced Qi regulation and promotion of intestinal microbiota balance.

Escherichia coli and LPS, gradually decrease. Considering that Wuqinxi can also improve risk symptoms such as obesity in middle-aged men with MS, it is speculated that optimizing the gut microbiota and its metabolites may be the key microbiological mechanism by which Wuqinxi promotes disease improvement.

5.3 Tai chi

Studies have shown that [60] 60 patients with ischemic stroke were divided into various groups, one of which participated in a 12-week Tai Chi Zhan Zhuang exercise program. Following the program, the results indicated that the gut microbiota of the patients in the experimental group had undergone significant changes. The number of beneficial bacteria originally present in the patients' intestines increased dramatically, while the number of harmful bacteria decreased significantly. This shift fully indicates that Tai Chi Zhan Zhuang exercise can positively regulate gut microbiota, thereby aiding in the improvement of the patients' intestinal micro-ecological environment.

Overall, TCM exercises, by regulating the spleen and stomach, unblocking meridians, and balancing Yin and Yang, can maintain the balance of the gut microbiota. This not only aids in improving gut health but also enhances overall health by regulating the immune system and metabolic activities (**Figure 5**).

6. The impact of psychological interventions on the gut microbiota

As previously mentioned, the gut microbiota is often referred to as the "second brain" of the human body. It not only affects our digestive system but also has a profound impact on emotions, mental health, and overall well-being through the gut-brain axis. The gut microbiota influences the secretion and regulation of neurotransmitters, immune responses, and endocrine hormones, making it an important factor in mental health that cannot be overlooked. The diversity and health status of the gut microbiota are closely related to psychological states such as anxiety and depression. If the gut microbiota is imbalanced, for instance, with a reduction in beneficial bacteria and an increase in harmful bacteria, it may exacerbate the occurrence of psychological problems. This relationship underscores the potential of the gut microbiota as a target for psychological interventions.

Psychological interventions, serving as a complementary and alternative therapy for enhancing emotional well-being, can directly or indirectly influence the gut microbiota by regulating stress. For instance, meditation and mindfulness therapy can lower stress levels, thereby reducing the damage inflicted on the gut microbiota. Studies have shown that psychological stress often triggers a "fight or flight" response, leading to the release of cortisol and catecholamines throughout the body, which can ultimately disturb the equilibrium of the gut microbiota. Meditation aids in regulating the stress response, thereby suppressing chronic inflammatory states and maintaining healthy gut barrier function [61]. As the classic saying goes, "Your mood determines your health." Psychological interventions not only regulate emotions but also improve the state of the gut microbiota.

6.1 Applications of psychological interventions in different diseases

6.1.1 Gastrointestinal diseases

The relationship between gastrointestinal diseases and mental disorders is increasingly drawing the attention of the academic community, particularly as symptoms of anxiety and depression often co-occur with IBD such as CD and irritable bowel syndrome (IBS). Dysbiosis of the gut microbiota may play a significant role in the connection between gastrointestinal diseases and symptoms of anxiety and depression.

Research has found that cognitive-behavioral therapy (CBT) can reduce the severity of IBS symptoms. Respondents to CBT exhibited increased levels of serotonin in their feces, an increase in the beneficial bacterial genus Clostridium, and a decrease in the harmful bacterial genus Bacteroides [62]. Psychological interventions can not only function independently in the treatment of gastrointestinal diseases but also synergistically with other therapies to enhance microbial dysbiosis in such conditions. A combination of two months of meditation and a vegan diet can significantly alter the composition of the microbiota, markedly increasing the abundance of beneficial bacterial genera and their metabolites, such as SCFAs and isovalerate [63].

6.1.2 Neurological and mental disorders

The gut microbiota communicates with the central nervous system via the "gut-brain axis," indicating that even minor changes in the central nervous system might be reflected through alterations in the gut microbiota. A randomized controlled trial on mindfulness-based practice revealed that [64] cognitive impairment in elderly patients with mild cognitive impairment was alleviated and correlated with the gut bacterial profile, where alterations in cognitive function led to changes in the abundance of specific gut microbes. Mindfulness also alleviates symptoms of anxiety and depression by influencing the gut microbiota. In individuals with high trait anxiety, mindfulness-based cognitive therapy (MBCT) has been found to reduce anxiety and

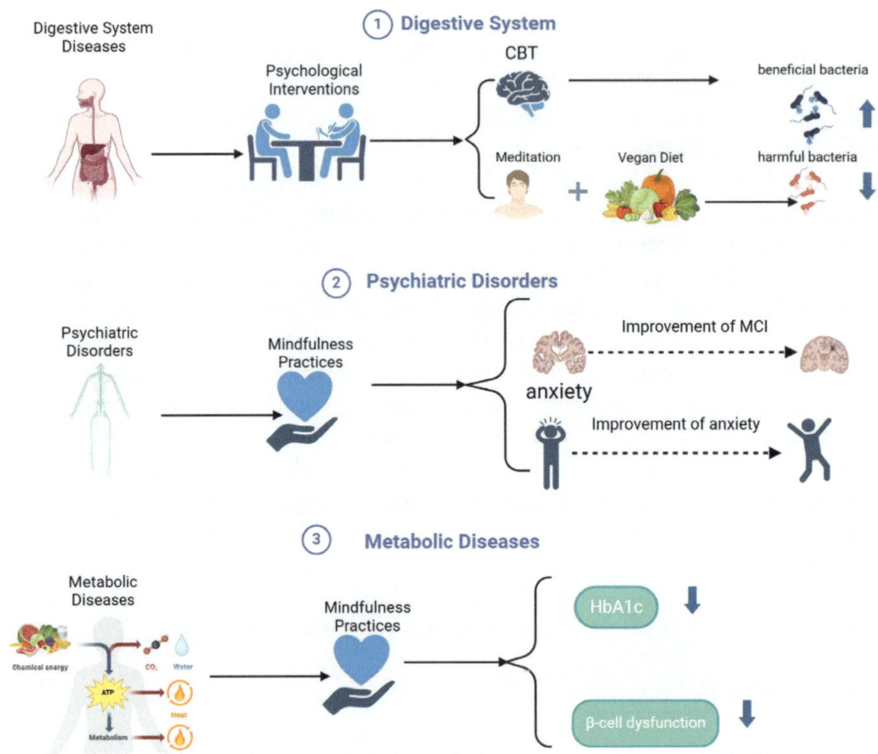

Figure 6.
Psychological interventions for disease treatment via modulation of the gut microbiota.

depression traits and increase the similarity of the gut microbiota to that of healthy controls. The reduction of Subdoligranulum after MBCT is closely associated with improved anxiety symptoms [65].

6.1.3 Metabolic disorders

Metabolic diseases are closely related to the gut microbiota, which can influence the host's metabolic processes through their metabolites. Firstly, gut microbes regulate immune responses and trigger chronic low-grade inflammation, leading to insulin resistance and fat accumulation. Secondly, gut microbiota metabolites, such as SCFAs, are involved in fat and carbohydrate metabolism, affecting the host's energy balance and lipid metabolism. Secondly, gut microbes engage with the gut-brain axis, affecting appetite and metabolic functions. A meta-analysis revealed that mindfulness can significantly lower glycated hemoglobin levels and enhance blood glucose control in individuals with diabetes. The gut microbiota may play a crucial role in the regulation of blood glucose levels through mindfulness [66]. Moreover, studies have also found significant differences in mindfulness levels among patients with type 2 diabetes at different levels of insulin resistance and β-cell function, indicating that abnormal insulin metabolism levels are associated with decreased mindfulness levels [67].

Overall, psychological interventions are closely related to the gut microbiota. On the one hand, alterations in the gut microbiota can impact psychological states, thereby influencing the onset of psychological disorders. On the other hand, chronic psychological stress can also lead to gastrointestinal dysfunction, which in turn affects the balance of the gut microbiota. Thus, psychological interventions, as a complementary therapy, have significant potential applications in improving the gut microbiota (**Figure 6**).

Acknowledgements

The authors would like to thank DeepSeek and Kimi AI tools for their technical support during the language correction process.

Conflict of interest

The authors declare no conflict of interest.

Author details

Mi Liu[1]*, Sihui Cao[1], Haihua Xie[1], Jia Jiang[1], Weiai Liu[2], Geshu Du[3], Shulin Xiong[2], Rong Luo[4] and Penghui Lu[1]

1 School of Acupuncture-moxibustion, Tuina and Rehabilitation, Hunan University of Chinese Medicine, Changsha, China

2 The Second Hospital of Hunan University of Chinese Medicine, Changsha, China

3 Changsha Hospital of Traditional Chinese Medicine Affiliated with Hunan University of Chinese Medicine, Changsha, China

4 The First Hospital of Hunan University of Chinese Medicine, Changsha, China

*Address all correspondence to: newmean@hnucm.edu.cn

References

[1] Zhang J, Wu Y, Tian Y, Xu H, Lin ZX, Xian YF. Chinese herbal medicine for the treatment of intestinal cancer: Preclinical studies and potential clinical applications. Molecular Cancer. 2024;**23**(1):217. DOI: 10.1186/s12943-024-02135-3

[2] Liu M, Shi W, Huang Y, Wu Y, Wu K. Intestinal flora: A new target for traditional Chinese medicine to improve lipid metabolism disorders. Frontiers in Pharmacology. 2023;**14**:1134430. DOI: 10.3389/fphar.2023.1134430

[3] Cao W, Wu J, Zhao X, Li Z, Yu J, Shao T, et al. Structural elucidation of an active polysaccharide from radix Puerariae lobatae and its protection against acute alcoholic liver disease. Carbohydrate Polymers. 2024;**325**:121565. DOI: 10.1016/j.carbpol.2023.121565

[4] Lu K, Li C, Men J, Xu B, Chen Y, Yan P, et al. Traditional Chinese medicine to improve immune imbalance of asthma: Focus on the adjustment of gut microbiota. Frontiers in Microbiology. 2024;**15**:1409128. DOI: 10.3389/fmicb.2024.1409128

[5] He Q, Han C, Huang L, Yang H, Hu J, Chen H, et al. Astragaloside IV alleviates mouse slow transit constipation by modulating gut microbiota profile and promoting butyric acid generation. Journal of Cellular and Molecular Medicine. 2020;**24**(16):9349-9361. DOI: 10.1111/jcmm.15586

[6] Liu X, Li M, Jian C, Wei F, Liu H, Li K, et al. Astragalus polysaccharide alleviates constipation in the elderly via modification of gut microbiota and fecal metabolism. Rejuvenation Research. 2022;**25**(6):275-290. DOI: 10.1089/rej.2022.0039

[7] Lin L, Jiang Y, Lin P, Ge L, Wan H, Dai S, et al. Classical famous prescription of Jichuan decoction improved loperamide-induced slow transit constipation in rats through the cAMP/PKA/AQPs signaling pathway and maintained inflammatory/intestinal flora homeostasis. Heliyon. 2024;**10**(1):e21870. DOI: 10.1016/j.heliyon.2023.e21870

[8] Wang XM, Lv LX, Qin YS, Zhang YZ, Yang N, Wu S, et al. Ji-Chuan decoction ameliorates slow transit constipation via regulation of intestinal glial cell apoptosis. World Journal of Gastroenterology. 2022;**28**(34):5007-5022. DOI: 10.3748/wjg.v28.i34.5007

[9] Gao Y, Wu X, Zhao N, Bai D. Scientific connotation of the compatibility of traditional Chinese medicine from the perspective of the intestinal flora. Frontiers in Pharmacology. 2023;**14**:1152858. DOI: 10.3389/fphar.2023.1152858

[10] Che Q, Luo T, Shi J, He Y, Xu DL. Mechanisms by which traditional Chinese medicines influence the intestinal flora and Intestinal barrier. Frontiers in Cellular and Infection Microbiology. 2022;**12**:863779. DOI: 10.3389/fcimb.2022.863779

[11] Hong J, Chen J, Kan J, Liu M, Yang D. Effects of acupuncture treatment in reducing sleep disorder and gut microbiota alterations in PCPA-induced insomnia mice. Evidence-based Complementary and Alternative Medicine. 2020;**2020**:3626120. DOI: 10.1155/2020/3626120

[12] Fang D, Shi D, Lv L, Gu S, Wu W, Chen Y, et al. Bifidobacterium pseudocatenulatum LI09 and Bifidobacterium catenulatum LI10

attenuate D-galactosamine-induced liver injury by modifying the gut microbiota. Scientific Reports. 2017;**7**(1):8770. DOI: 10.1038/s41598-017-09395-8

[13] Walker AW, Sanderson JD, Churcher C, Parkes GC, Hudspith BN, Rayment N, et al. High-throughput clone library analysis of the mucosa-associated microbiota reveals dysbiosis and differences between inflamed and non-inflamed regions of the intestine in inflammatory bowel disease. BMC Microbiology. 2011;**11**:7. DOI: 10.1186/1471-2180-11-7

[14] Wei D, Xie L, Zhuang Z, Zhao N, Huang B, Tang Y, et al. Gut microbiota: A new strategy to study the mechanism of electroacupuncture and moxibustion in treating ulcerative colitis. Evidence-based Complementary and Alternative Medicine. 2019;**2019**:9730176. DOI: 10.1155/2019/9730176

[15] Tuoyire DA, McNair S, Debrah SA, Duda RB. Perception of risk for hypertension and overweight/obesity in Cape Coast, Ghana. Ghana Medical Journal. 2018;**52**(3):140-146. DOI: 10.4314/gmj.v52i3.6

[16] Si YC, Miao WN, He JY, Chen L, Wang YL, Ding WJ. Regulating gut flora dysbiosis in obese mice by electroacupuncture. The American Journal of Chinese Medicine. 2018;**46**(7):1-17. DOI: 10.1142/S0192415X18500763

[17] Li Y, Wang S, Bai Y, Fan MM. Exploration of the application of acupuncture in the treatment of depression in Parkinson's disease based on the theory of gut microbiota. Journal of Clinical Acupuncture and Moxibustion. 2018;**34**(10):79-82. DOI: 10.3969/j.issn.1005-0779.2018.10.023

[18] Xie WS, Wei LC, Wang JZ, Zhang F. Efficacy of acupuncture combined with traditional Chinese medicine in the treatment of ulcerative colitis and its impact on related indicators. Shanghai Journal of Acupuncture and Moxibustion. 2019;**38**(04):378-383. DOI: 10.13460/j.issn.1005-0957.2019.04.0378

[19] Li XL, Cai JZ, Lin Y, Liu XN. Study on the effect of acupuncture therapy in regulating the gut microbiota structure and promoting intestinal tight junctions in rats with IBS-D. Journal of Guangzhou University of Traditional Chinese Medicine. 2019;**36**(07):1022-1028. DOI: 10.13359/j.cnki.gzxbtcm.2019.07.019

[20] Wang LJ, Xue T, Wu YQ, Zhao JY, Wang TN, Li JT, et al. Effects of acupuncture on the gut microbiota of rat models with stress-induced gastric ulcer. China Acupuncture Moxibustion. 2020;**40**(05):526-532. DOI: 10.13703/j.0255-2930.20191121-0002

[21] Chang XR, Liu M, Yan J, Yi SX, Yue ZH, Zhang GS, et al. Study on the mechanism and patterns of the warming and nourishing effects of moxibustion. World Chinese Medicine Forum. 2013;**8**:875-879. DOI: 10.3969/j.issn.1673-7202.2013.08.008

[22] Shimma S, Sugiura Y, Hayasaka T, Hoshikawa Y, Noda T, Setou M. MALDI-based imaging mass spectrometry revealed abnormal distribution of phospholipids in colon cancer liver metastasis. Journal of Chromatography. B, Analytical Technologies in the Biomedical and Life Sciences. 2007;**855**(1):98-103. DOI: 10.1016/j.jchromb.2007.02.037

[23] Feng L, Ding MH, Shen Q, Sun QH, Shi SL. Study on the factors affecting the transdermal permeability of rhubarb acid at the shenque acupoint: A preliminary mechanism exploration. Chinese

Medicinal Materials. 2016;**41**(08):1546-1552. DOI: 10.4268/cjcmm20160830

[24] Jiang Y, Dai WJ, Jiang JF. Differences in microcirculation perfusion of common abdominal acupoint areas and the effects of moxibustion at different temperatures. Journal of Nanjing University of Traditional Chinese Medicine. 2017;**33**(04):339-343. DOI: 10.14148/j.issn.1672-0482.2017.0339

[25] Yu ZR, Xian J, Wang L, Yu HJ. The effect of umbilical moxibustion on gut microbiota in patients with subthreshold depression. China Acupuncture Moxibustion. 2024;**44**(08):881-888+918. DOI: 10.13703/j.0255-2930.20231019-0003

[26] Jing C, Xiao N, Yu HJ, Zhang CY, Zhang X, Tan QW. The effect of umbilical moxibustion on regulating phlegm-damp constitution and its impact on the gut microbiota of subjects. China Acupuncture Moxibustion. 2021;**41**(12):1360-1364+1382. DOI: 10.13703/j.0255-2930.20210430-k0001

[27] Liu J, An N, Ma C, Li X, Zhang J, Zhu W, et al. Correlation analysis of intestinal flora with hypertension. Experimental and Therapeutic Medicine. 2018;**16**(3):2325-2330. DOI: 10.3892/etm.2018.6500

[28] Zhu JY, Zheng T, Yang Y, Tan ZJ. Research progress on the correlation between intestinal microbiota and its metabolites and hypertension. Chinese Journal of Microecology. 2022;**34**(03):347-351. DOI: 10.13381/j.cnki.cjm.202203020

[29] Huang XJ, Yang J, Lei YQM, Zeng ZJ, Jin ZH. Research progress on novel auxiliary antihypertensive modes targeting intestinal microbiota. Chinese General Practice. 2022;**25**(18):2297-2302. DOI: 10.12114/j.issn.1007-9572.2022.02.023

[30] Jiang XY. Clinical observation on the treatment of essential hypertension based on real-world heat-sensitive umbilical cord moxibustion [thesis]. Nanchang: Jiangxi University of Traditional Chinese Medicine; 2024

[31] Song JW, Liao YJ, Ma YJ, Rao ZH, Yue ZH. Research progress on the mechanism of vascular endothelial function in atherosclerosis protected by herb cake moxibustion. China Journal of Traditional Chinese Medicine and Pharmacy. 2022;**37**(09):5273-5276

[32] Liao YS. Clinical observation on the treatment of diarrheal irritable bowel syndrome with medicine cake moxibustion [thesis]. Nanjing: Nanjing University of Traditional Chinese Medicine; 2022

[33] Lin YF, Jin XQ, Zhu JF, Chen YD, Sheng JL, He JJ, et al. Treatment of chronic fatigue syndrome with ginger moxibustion and its effect on intestinal microbiota. China Acupuncture Moxibustion. 2021;**41**(03):269-274. DOI: 10.13703/j.0255-2930.20200210-k0001

[34] Fu X, Liu Z, Zhu C, Mou H, Kong Q. Nondigestible carbohydrates, butyrate, and butyrate-producing bacteria. Critical Reviews in Food Science and Nutrition. 2019;**59**(sup1):S130-SS52. DOI: 10.1080/10408398.2018.1542587

[35] Yu ZT, Hang SQ, Yao W, Zhu WY. Research progress on intestinal butyric-producing bacteria and their butyric acid production mechanism. World Chinese Journal of Digestology. 2006;**14**(25):2531-2534. DOI: 10.3969/j.issn.1009-3079.2006.25.012

[36] Chen RZ, Li PF, Hu YQ, Xu Y, Zhang LL, Yuan JL, et al. The role of

intestinal microbiota in the metabolism of traditional Chinese medicine. Chinese Journal of Microecology. 2018;**30**(8):990-992,3. DOI: 10.13381/j.cnki.cjm.201808029

[37] Feng Y, Wang Y, Wang P, Huang Y, Wang F. Short-chain fatty acids manifest stimulative and protective effects on intestinal barrier function through the inhibition of NLRP3 inflammasome and autophagy. Cellular Physiology and Biochemistry. 2018;**49**(1):190-205. DOI: 10.1159/000492853

[38] Chen RX. Heat-sensitive moxibustion: The inheritance, innovation and development of traditional Chinese medicine moxibustion. China Acupuncture Moxibustion. 2023;**43**(04):483-488. DOI: 10.13703/j.0255-2930.20220625-k0004

[39] Pan H, Zhang SM, Wang X, Jiang DD. Effect of Qi Chang Xuexing Tang retention enema combined with heat-sensitive moxibustion in radical gastric cancer resection. Journal of Practical Medicine. 2023;**39**(16):2130-2135. DOI: 10.3969/j.issn.1006-5725.2023.16.020

[40] Huang Y, Ma Q, He J, Liang X, Mai Q, Luo H, et al. Abdominal massage alleviates functional diarrhea in immature rats via modulation of intestinal microbiota and tight junction protein. Frontiers in Pediatrics. 2022;**10**:922799. DOI: 10.3389/fped.2022.922799

[41] Gao H, Zhang X, He W, Zhao X, Han J, Li D, et al. To study the intervention mechanism of pediatric massage on intestinal flora and host metabolism in children with anorexia. Medicine (Baltimore). 2020;**99**(47):e23349. DOI: 10.1097/MD.0000000000023349

[42] Feng W, Wang Y, Zhang Y. Effects of abdominal Tuina on “microbiota-brain-gut axis” signal pathway in STC rats. Liaoning Journal of Traditional Chinese Medicine. 2022;**49**(05):200-202. DOI: 10.13192/j.issn.1000-1719.2022.05.055

[43] Chen H, Tan PS, Li CP, Chen BZ, Xu YQ, He YQ, et al. Acupoint massage therapy alters the composition of gut microbiome in functional constipation patients. Evidence-based Complementary and Alternative Medicine. 2021;**2021**:1416236. DOI: 10.1155/2021/1416236

[44] Xue TJ. Effects of abdominal Tuina therapy on intestinal flora and lipid metabolism in type 2 diabetic rats [thesis]. Beijing: Beijing University of Chinese Medicine; 2019

[45] Xie Y, Huan MT, Sang JJ, Luo SS, Kong XT, Xie ZY, et al. Clinical effect of abdominal massage therapy on blood glucose and intestinal microbiota in patients with type 2 diabetes. Oxidative Medicine and Cellular Longevity. 2022;**2022**:2286598. DOI: 10.1155/2022/2286598

[46] Li T, Feng X, Zhi H, Huang W, Li J, Li W. Effects of tuina targeting different body parts on the behaviors and gut microflora of autistic spectrum disorder rat models. Digital Chinese Medicine. 2024;**7**(1):90-100. DOI: 10.1016/j.dcmed.2024.04.010

[47] Fu BL.Observation of the clinical efficacy of vibration massage therapy in children with autism and changes in intestinal flora [thesis]. Beijing: Beijing University of Chinese Medicine; 2021

[48] Chen ML, Yi L, Zhang Y, Zhou X, Ran L, Yang J, et al. Resveratrol attenuates trimethylamine-N-oxide (TMAO)-induced atherosclerosis

by regulating TMAO synthesis and bile acid metabolism via remodeling of the gut microbiota. MBio. 2016;**7**(2):e02210-e02215. DOI: 10.1128/mBio.02210-15

[49] Sonnenburg ED, Smits SA, Tikhonov M, Higginbottom SK, Wingreen NS, Sonnenburg JL. Diet-induced extinctions in the gut microbiota compound over generations. Nature. 2016;**529**(7585):212-215. DOI: 10.1038/nature16504

[50] Rahman MM, Islam F, Or-Rashid MH, Mamun AA, Rahaman MS, Islam MM, et al. The gut microbiota (microbiome) in cardiovascular disease and its therapeutic regulation. Frontiers in Cellular and Infection Microbiology. 2022;**12**:903570. DOI: 10.3389/fcimb.2022.903570

[51] Lopez-Garcia E, Rodriguez-Artalejo F, Li TY, Fung TT, Li S, Willett WC, et al. The Mediterranean-style dietary pattern and mortality among men and women with cardiovascular disease. The American Journal of Clinical Nutrition. 2014;**99**(1):172-180. DOI: 10.3945/ajcn.113.068106

[52] Ferrere G, Tidjani Alou M, Liu P, Goubet AG, Fidelle M, Kepp O, et al. Ketogenic diet and ketone bodies enhance the anticancer effects of PD-1 blockade. JCI Insight. 2021;**6**(2):e145207. DOI: 10.1172/jci.insight.145207

[53] Giampazolias E, Pereira da Costa M, Lam KC, Lim KHJ, Cardoso A, Piot C, et al. Vitamin D regulates microbiome-dependent cancer immunity. Science. 2024;**384**(6694):428-437. DOI: 10.1126/science.adh7954

[54] Garcia-Penas JJ. Autism spectrum disorder and epilepsy: The role of ketogenic diet. Revista de Neurologia. 2016;**62**(Suppl. 1):S73-S78

[55] Liu Y, Yang TY, Qi X, Tang XY. Mechanisms and research advances in the regulation of depression by gut microbiota via the gut-brain Axis. Journal of Mudanjiang Medical University. 2024;**45**(06):135-139. DOI: 10.13799/j.cnki.mdjyxyxb.2024.06.022

[56] Provensi G, Schmidt SD, Boehme M, Bastiaanssen TFS, Rani B, Costa A, et al. Preventing adolescent stress-induced cognitive and microbiome changes by diet. Proceedings of the National Academy of Sciences of the United States of America. 2019;**116**(19):9644-9651. DOI: 10.1073/pnas.1820832116

[57] Gram-Kampmann EM, Hansen CD, Hugger MB, Jensen JM, Brond JC, Hermann AP, et al. Effects of a 6-month, low-carbohydrate diet on glycaemic control, body composition, and cardiovascular risk factors in patients with type 2 diabetes: An open-label randomized controlled trial. Diabetes, Obesity & Metabolism. 2022;**24**(4):693-703. DOI: 10.1111/dom.14633

[58] Marasco G, Cirota GG, Rossini B, Lungaro L, Di Biase AR, Colecchia A, et al. Probiotics, prebiotics and other dietary supplements for gut microbiota modulation in celiac disease patients. Nutrients. 2020;**12**(9):2674. DOI: 10.3390/nu12092674

[59] Sun HM, Zhong YP. Research on the intervention of Wuqinxi on intestinal Flora and its metabolites of middle-aged Men with metabolic syndrome. Journal of Chengdu Sport University. 2019;**45**(03):81-87. DOI: 10.15942/j.jcsu.2019.03.013

[60] Wang F. A study on the effects of Tai Chi stance training on motor function impairment and gut microbiota in patients with ischemic stroke [thesis]. Heilongjiang: Harbin Sport University; 2023

[61] Househam AM, Peterson CT, Mills PJ, Chopra D. The effects of stress and meditation on the immune system, human microbiota, and epigenetics. Advances in Mind-Body Medicine. 2017;**31**(4):10-25

[62] Jacobs JP, Gupta A, Bhatt RR, Brawer J, Gao K, Tillisch K, et al. Cognitive behavioral therapy for irritable bowel syndrome induces bidirectional alterations in the brain-gut-microbiome axis associated with gastrointestinal symptom improvement. Microbiome. 2021;**9**(1):236. DOI: 10.1186/s40168-021-01188-6

[63] Raman M, Vishnubhotla R, Ramay HR, Gonçalves MCB, Shin AS, Pawale D, et al. Isha yoga practices, vegan diet, and participation in Samyama meditation retreat: Impact on the gut microbiome & metabolome - a non-randomized trial. BMC Complementary Medicine and Therapies. 2023;**23**(1):107. DOI: 10.1186/s12906-023-03935-8

[64] Khine WWT, Voong ML, Ng TKS, Feng L, Rane GA, Kumar AP, et al. Mental awareness improved mild cognitive impairment and modulated gut microbiome. Aging (Albany NY). 2020;**12**(23):24371-24393. DOI: 10.18632/aging.202277

[65] Wang Z, Liu S, Xu X, Xiao Y, Yang M, Zhao X, et al. Gut microbiota associated with effectiveness and responsiveness to mindfulness-based cognitive therapy in improving trait anxiety. Frontiers in Cellular and Infection Microbiology. 2022;**12**:719829. DOI: 10.3389/fcimb.2022.719829

[66] Hamasaki H. The effects of mindfulness on glycemic control in people with diabetes: An overview of systematic reviews and meta-analyses. Medicines (Basel). 2023;**10**(9):53. DOI: 10.3390/medicines10090053

[67] Zhang X, Huang R, Li J, Yang M, Zhang D, Liu C, et al. Abnormal insulin metabolism and decreased levels of mindfulness in type 2 diabetes mellitus. Diabetology and Metabolic Syndrome. 2025;**17**(1):32. DOI: 10.1186/s13098-025-01594-y

Chapter 6

Correlation of High Gut Microbiota Archaea Methanogenesis with Health Characteristics of Humans and Animals

Oleg Medvedev, Oxana Povarova, Nikita Fadeev, Ghizal Fatima, Ustin Zolotikov and Stepan Toshchakov

Abstract

A review of the current literature on gut microbiota biomarkers reveals a very significant imbalance in research focus. Most studies focus on the taxonomic composition of bacterial microbiota in various clinical conditions, such as obesity, gastrointestinal tract diseases, cardiovascular diseases, and type 2 diabetes (T2DM). In contrast, there are relatively few studies examining methanogenic archaea. This discrepancy may stem from a long-held belief that archaea are not pathogenic microorganisms in humans, coupled with the limited sensitivity of the widely used 16S rRNA method for detecting methanogenic archaea. Several publications highlight the functional differences between predominant hydrogen producers and methane producers. High levels of methane production by microbiota are correlated with obesity, constipation, lower levels of short-chain fatty acids in the intestinal lumen, immune changes, unhealthy aging, and carcinogenesis. Nutritional factors have primarily been investigated to reduce methanogenesis and archaea abundance in livestock, aiming to mitigate ecological issues like global warming. In humans, evidence suggests that certain statins and antibiotics, as well as low FODMAP diets and probiotics, can decrease methane production. Findings from ruminant livestock studies on inhibiting methane production could hold promise for clinical evaluation. We propose that user-friendly, non-invasive, and affordable methods are needed to screen methane-producing individuals. Such methods would facilitate the development of personalized nutritional recommendations and help prevent the onset of various non-communicable diseases.

Keywords: hydrogen, methane, gut microbiota, methanogenic archaea, next-generation sequencing

1. Introduction

The human gastrointestinal tract is home to the largest microbial community in the body, composed of trillions of microbes known as gut microbiota. This normal flora plays a crucial role in various physiological functions, such as enhancing host immunity, aiding in nutrient absorption, and protecting the body against pathogenic microorganisms. Within the gut, chemical interactions among the microbiota produce gases, including carbon dioxide (CO_2), hydrogen (H_2), methane (CH_4), and hydrogen sulfide (H_2S), along with a variety of trace gases. Profiling these intestinal gaseous biomarkers and examining their interrelations, as well as their effects on different organs and systems within the human body, can provide insights into the products and functions of gut microbiota and their influence on human health [1, 2].

This review focuses on the differences in metabolic activity—specifically the production of H_2, CH_4, trimethylamine (TMA), trimethylamine N-oxide (TMAO), and short-chain fatty acids as well as the associated risks of non-communicable diseases (NCDs) and longevity between predominant producers of H_2 and CH_4. These classifications can be determined through breath tests or by analyzing the composition of intestinal gas. Approximately 65% of all bacteria in the gut microbiome possess hydrogenase genes, enabling them to produce hydrogen, while methane is exclusively generated by archaea that utilize H_2 and carbon- or methyl-containing molecules. This distinction explains the reciprocal relationship between H_2 and CH_4 levels in the intestinal lumen and in exhaled air, which essentially reflects the composition of gaseous gut content.

Patients diagnosed with CH_4-dominant intestinal bacterial overgrowth often report symptoms such as bloating, abdominal distension, and constipation [3, 4]. Preliminary studies suggest that this diagnosis can help guide optimal antibiotic treatment choices for clinical benefit [5–7]. Currently, breath tests are the primary method for identifying CH_4 producers; however, these tests have limitations in sensitivity and logistical challenges. Moreover, there is significant discordance between CH_4 detection via breath tests and colonic CH_4 concentration, particularly in patients with irritable bowel syndrome (IBS) [8, 9]. Presently, the 16S rRNA–based next-generation sequencing (NGS) analysis of the microbiome is an increasingly prevalent approach for investigating disease-related alterations in gut flora. This is primarily attributable to the rapid reduction in cost and the development of streamlined bioinformatic analysis pipelines. However, this approach seems to have limited sensitivity regarding reliable detection of archaea abundances in the human gut, especially in large-scale screening studies, where comparatively low sequencing depth is utilized.

Understanding the physiology and pathogenesis of various diseases and diagnostics can be greatly enhanced by the composition of intestinal gases [10]. The most steep oxygen gradient in the body can be achieved through the unique anatomic and physiologic features of healthy intestines. The gradient is utilized by the intestines and microbiome to create redox effectors like nitric oxide, hydrogen sulfide, and reactive oxygen species that come from both hosts and bacterial origin (**Figure 1**) [12].

Diet has a significant impact on the composition of intestinal gas profiles among individuals [13]. It has been reported that a variety of diets can cause a rapid and reproducible modification of the intestinal microbiome [14–16]. Some diseases occur as a result of imbalances in gaseous syntheses. Elevated hydrogen (H2) levels may indicate lactose and fructose intolerance [17] as well as a small intestine bacterial overgrowth [18]. Taking into account hydrogen's antioxidant, anti-inflammatory, and anti-apoptotic effects, an increase in hydrogen levels can lead to a high level of antioxidant protection for a living organism. Japanese centenarians with a high level of hydrogen in their exhaled air are a confirmation of this fact [19]. By comprehending the composition of physiological intestinal gases and their pathological variations, one can gain valuable knowledge about the functionality of the microbiota,

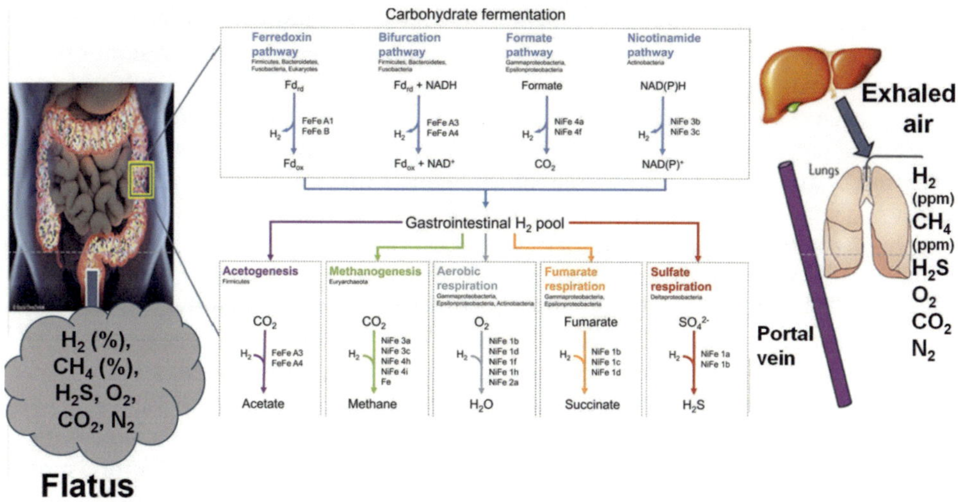

Figure 1.
Production and kinetics of gas biomarkers of the intestinal microbiota. This figure was adapted from Ref. [11].

the pathophysiology of various intestinal diseases, and how different types of diets affect the production of intestinal gases.

Most published clinical and experimental studies concern measurements of hydrogen and methane in exhaled air [4, 18, 19]. According to a published metaanalysis of clinical studies conducted, hydrogen was measured in exhaled air in 20.5 % of the people [20]. Methane was detected in flatus in an average of 53% of people in the general population (**Table 1**) [21].

A special breath test determines the level of methane in exhaled air, which reflects the level of methanogens in the intestine. It has been experimentally established that detection of CH_4 in exhaled air at a concentration of >1 ppm corresponds to a minimum concentration of methanogens in the intestine of 10^7–10^8/g feces [22]. Using this test, animals and people can be divided into methanogens and non-methanogens.

The current recommendation is to use fasting single methane measurements (SMM) to diagnose intestinal methanogen overgrowth (IMO) and monitor treatment response in individuals with IMO [23]. With a cutoff of SMM $\geq$ 10 ppm, SMM was highly accurate in diagnosing IMO on the glucose and lactulose breath tests, with a sensitivity of 86% and specificity of 100% and was linked to constipation ($p = 0.008$). SMM remained constant for 14 weeks without any treatment ($P = 0.45$), but antibiotics caused a decrease in SMM within 2 days ($P < 0.0001$). A positive correlation between SMM and the load of stool *Methanobrevibacter smithii* ($R = 0.65$, $P < 0.0001$) was observed.

Methanogenic archaea are a group of archaea that reside in anaerobic conditions within the digestive system. Archaea represent the domain of living organisms

Gas species	% of participants present	Mean % concentration ± SD	Range (%)
H_2	100%	2.9 ± 0.7	0.17–49.0
CH_4	53.6%	14.4 ± 3.7	0–30.0

This table was adapted from Ref. [21].

Table 1.
Composition of colonic gases (H_2 and CH_4) in healthy participants with a regular diet.

according to the three-domain system of Carl Woese, occupying an intermediate position between prokaryotes and eukaryotes [24]. These are single-celled microorganisms with similar structural elements characteristic of prokaryotes and eukaryotes. Initially, archaea were considered extremophiles living in extreme conditions. In recent decades, a sufficient amount of data has been accumulated indicating the participation of archaea in the ecosystem of living organisms, including mammals. The conducted studies showed that archaea have distinct compositions depending on their location in the human body [25]. Methanogenic archaea is the final electron acceptors in the food chain in the anaerobic conditions of the gastrointestinal tract, converting energy with the formation of methane. The synthesis of ATP by methanogenesis is crucial for archaea's growth in anaerobic conditions. Currently, methanogenesis can be divided into three directions depending on the sources in association with the hosts: hydrogenotrophic, methylotrophic, and acetoclastic [26].

The most common direction of methanogenesis among archaea associated with humans is hydrogenotrophic, using hydrogen (four moles) to reduce CO_2 (one mole) for the synthesis of one molecule of methane [27]. *Methanobrevibacter* is the most common type of archaea among mammals, including humans, carrying out this type of methanogenesis (**Figure 2**). This pathway is aimed at reducing the excess hydrogen formed during the life of bacteria, which creates favorable conditions for symbiosis of archaea with other representatives of the intestinal microbiota.

Methylotrophic methanogenesis is accompanied by the reduction of methylated compounds (methanol, trimethylamine, and its precursors) to methane [29]. This pathway is characteristic of *Methanosphaera stadtmanae* and representatives of *Methanomassiliicoccales*. Along with the hydrogenotrophic pathway, it is one of the most common pathways of host-associated strains (**Figure 3**). A feature of this methanogenesis pathway is the utilization of trimethylamine with subsequent reduction of the level of proatherogenic TMAO [30, 31]. This fact is actively discussed in the literature as a promising direction for the use of archaea as archaeobiotics in the treatment of metabolic diseases [32].

Acetoclastic methanogenesis is the most common methanogenesis pathway in the environment and less common among living organisms [33]. It is aimed at the breakdown of acetate with the formation of methane and is represented by archaea belonging to the genus *Methanosarcinales*.

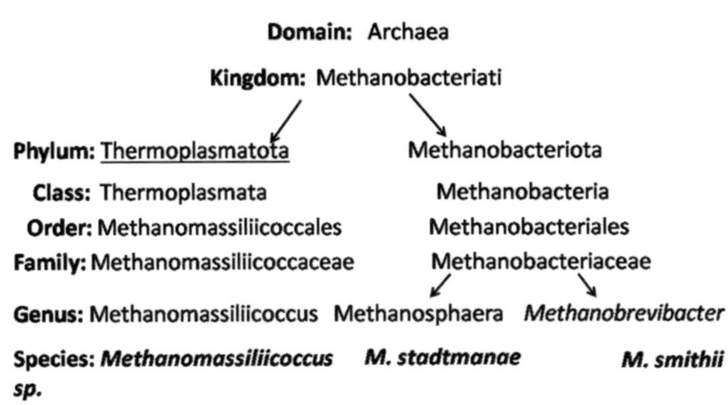

Figure 2.
Taxonomy of Methanobrevibacter smithii, M. stadtmanae, *and* Methanomassiliicoccus sp. *[28].*

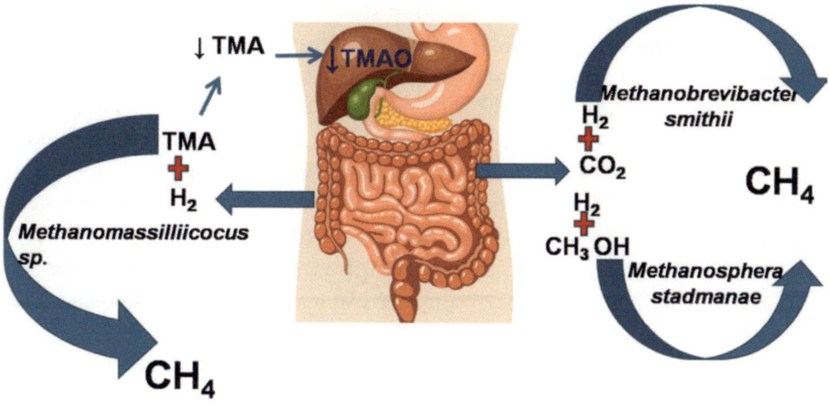

Figure 3.
Variants of methane generation by intestinal archaea. Explanations are provided in the main text.

2. Archaea detection

Molecular genetics technologies, including PCR and sequencing, have made it possible to detect archaea more often, even where their presence was previously technically impossible.

The utilization of 16S rRNA–based next-generation sequencing (NGS) for microbiome analysis has witnessed a marked increase in recent times, with this approach gaining prominence as a means to investigate disease-related alterations in gut microbiota. This growth in popularity can be attributed to a confluence of factors, including the substantial reduction in costs and the development of sophisticated and user-friendly bioinformatic analysis tools. However, this method appears to have limited sensitivity when it comes to reliably detecting archaeal populations in the human gut, particularly in large-scale screening studies where relatively low sequencing depth is employed. This limitation might stem from several potential factors, such as biases caused by inefficient cell lysis of methanogenic archaea, insufficient specificity of universal primers, and the lower number of ribosomal operons in methanogenic archaea compared to other dominant gastrointestinal taxa. The following section will delve into these potential factors in greater detail, exploring their potential implications for underestimating methanogenic archaea abundances.

2.1 Metagenomic DNA extraction protocol-related biases

The process of identifying methanogens in environmental samples can be hindered by limitations in extraction methods, potentially leading to an underestimation of their diversity and abundance. Methanogens' cell walls lack peptidoglycan but are reinforced by pseudomurein or proteinaceous S-layers, which resist chemical lysis agents like lysozyme or proteinase K used in standard protocols [34]. Furthermore, mechanical lysis techniques, which rely on a bead-beating step and are regarded as the "gold standard" for microbiome studies, are often applied inconsistently or inadequately optimized in many commercial kits. This results in preferential lysis of Gram-negative bacteria over methanogens and other resilient microorganisms [35, 36].

Several research groups, focusing on methanogenesis in human gut, develop custom protocols for methanogen identification. Thus, Dridi et al. employed a double mechanical cell lysis procedure involving glass beads. Between two lysis steps, they subjected the samples to an overnight incubation with proteinase K. This modification of the protocol led to a 100- to 1000-fold increase in the number of detectable copies of 16S rRNA/rpoB genes of methanogenic archaea per 1 gram of feces [37]. A recent work by Cisek and co-authors, aimed at enhancing the efficiency of the protocol for extracting archaeal DNA from fecal samples, has demonstrated that the incorporation of an additional ultrasound treatment step, employing an immersion probe sonicator, results in a substantial increase (135x compared to non-mechanically lysed control) in the yield of methanogenic archaeal DNA [38].

These findings underscore the critical role of extraction protocols in accurately assessing methanogen populations and highlight the necessity of method optimization to circumvent biases in archaea community analyses. The most evident approach to address method-dependent variation in DNA extraction efficiency is to employ synthetic microbial communities ('mock communities') containing predetermined ratios of cells of different bacterial taxa to validate extraction protocols and further sample preparation [39]. The incorporation of synthetic microbial communities into routine studies, conducted in parallel with experimental samples, has the potential to facilitate the timely identification of emerging deviations. This, in turn, could enable the refinement of extraction protocols, thereby ensuring the accuracy and reliability of methanogen population assessments.

A recent extensive benchmarking study using mock communities, including methanogenic archaea *Methanobrevibacter smithii*, was conducted by an interlaboratory consortium within the MOSAIC Standards Challenge. The study's findings indicated that the most effective sample preparation protocols, as indicated by the capacity to accurately match the detectable representation of *Methanobrevibacter smithii* to a known concentration within a mock community, were those employing at least threefold mechanical lysis utilizing zirconia beads [40].

Taken together, these facts emphasize the importance of detailed cell lysis protocol selection in the isolation of methanogenic archaea DNA in gut microbiomes and point to a possible reason for the low detection rate of methanogens in publicly available data.

2.2 Biases related to hypervariable region and primer pair choice

The selection of the hypervariable region of the 16S rRNA gene, in conjunction with a specific primer pair for this region, exerts a substantial influence on the observable composition of the microbial community [41, 42]. A comprehensive review of the extant literature, encompassing the findings of several large-scale projects, including the Earth Microbiome Project and Human Metagenome Project, reveals a consensus among researchers worldwide regarding the efficacy of the V4 or V3/V4 hypervariable region [43, 44]. The rationale behind this consensus lies in the superior coverage of diverse taxa afforded by primers selected from the conserved regions flanking these regions, complemented by the high discriminatory power of variable sequences within these regions [45, 46]. It is important to acknowledge that the selection of relatively short amplicons is driven by the constrained length of paired-end reads in second-generation sequencing systems, such as Illumina, Ion Torrent, or DNBSeq, which typically does not exceed 300 nucleotides. The amplicon lengths of 460 and 280 nucleotides at the V3-V4 and V4 regions, respectively, ensure complete overlap with paired-end reads, facilitating efficient and reliable read merging. It is important to consider, however, that the V3-V4 region exhibits variation in length between bacteria and archaea. The archaeal V3-V4 region is comparatively shorter than the bacterial region, resulting in an overestimation of archaea abundances in microbiomes. This overestimation can be attributed to the increased PCR efficiency observed for shorter fragments [47].

Presently, third-generation sequencing technologies are undergoing active development, enabling the production of long reads that fully overlap the sequence of the 16S rRNA gene. This advancement addresses the challenge of selecting a hypervariable region. However, these technologies currently incur a higher cost compared to highly parallel sequencing (second generation), and their relatively high read error rate limits the attainment of ultrahigh resolution at the species or strain level [48].

The efficacy of specific taxa detection is contingent upon the selection of appropriate primers for the targeted hypervariable regions. Important parameters include the primer position and the utilization of ambiguous bases. As demonstrated by Takahashi et al., the incorporation of a single ambiguity in the prokaryotic universal primer Uni340F [49] results in a nearly twofold increase in the detection of methanogenic archaea [50]. **Table 2** offers a synopsis of the *in silico* annealing analysis of various primers to the V4 hypervariable region on different taxa of the order *Methanobacteriales* performed with Silva TestPrime web tool [55]. The results of the analysis demonstrate that it is virtually impossible to select a primer that demonstrates a significant improvement in coverage of individual taxonomic groups without compromising other taxa. To illustrate this point, the primers described by Merkel et al. [54] exhibit a substantial increase in coverage of the family *Methanomethylophilaceae* of the order *Methanomassiliicoccales* (from 79.8 to 94.5%), while concomitantly resulting in a slight decrease in coverage of the family *Methanobacteriaceae* of the order *Methanobacteriales* (from 90.6 to 90.4%). These results underscore the notion that primer biases, most notably those pertaining to understudied archaeal lineages, have the potential to contribute to an underestimation of the abundance of certain taxa in 16S rRNA microbiota profiling studies.

Primer name and reference*	EMP primers [51]	Parada et al. [52]	Walters et al. [53]	Merkel et al. [54]
Methanomassiliicoccales	**83.8**	**84.0**	**83.8**	**91.2**
Unknown family	90.7	91.9	90.7	91.9
Methanomassiliicoccaceae	96.8	96.8	96.8	96.8
Methanomethylophilaceae	79.8	79.8	79.8	94.5
Methanobacteriales	**90.5**	**90.5**	**90.5**	**90.4**
Methanobacteriaceae	90.6	90.6	90.6	90.4
Methanothermaceae	100	100	100	100
Methanothermobacteriaceae	89.4	89.4	89.4	89.4

Primer sequences: EMP primers – GTGCCAGCMGCCGCGGTAA and GGACTACHVGGGTWTCTAAT; Parada primers – GTGYCAGCMGCCGCGGTAA and GGACTACHVGGGTWTCTAAT; Walters primers – GTGCCAGCMGCCGCGGTAA and GGACTACNVGGGTWTCTAAT; Merkel primers – GTGBCAGCMGCCGCGGTAA and GACTACNVGGGTMTCTAATCC.

Table 2.
Coverage of different taxa of methanogenic archaea by the most common primer pairs to the V4 hypervariable region.

In summary, although exhibiting certain biases and with ample opportunity for enhancement, the utilization of universally applicable 16S rRNA gene primers (e.g., 515F/806R) has served as a foundational element in microbiota profiling. However, these primers have been observed to demonstrate a tendency to exhibit bias against archaeal communities, including methanogens. These "universal" primers target regions that are conserved across bacteria; however, they may not efficiently amplify archaeal sequences due to mismatches in primer-binding sites. This can result in underrepresentation of methanogenic archaea in microbiome datasets [53, 56]. The employment of archaeal-specific 16S rRNA primers, such as Arch344F/519R or A571F/UA1204R, has been demonstrated to enhance detection sensitivity by targeting hypervariable regions that are optimized for archaeal diversity. The utilization of these primers has been shown to reduce bacterial cross-amplification and enhance the recovery of low-abundance methanogens, including genera such as *Methanobrevibacter* and *Methanosphaera*, which are of significant importance within the GIT environment [57, 58]. A series of comparative studies have been conducted to ascertain the impact of utilizing universal versus archaea-specific primers. The findings of these studies have demonstrated that the latter significantly enhance the resolution of methanogen communities, resulting in the revelation of higher phylogenetic diversity and abundance [59, 60]. In particular, the recent work of Thomas and colleagues employed nested archaea-specific PCR with a primer pair that was specific to a large fragment of the 16S rRNA gene, encompassing the V3-V7 hypervariable regions, and a second primer pair that was specific to the V4-V5 regions. This approach enhanced the detection of archaea in samples of the gut microbiota of animals almost twice and significantly increased the diversity of gut-associated archaea in 16S-profiling experiments [60]. Therefore, the targeted approach is of particular value for the study of methane-producing microbiomes, where the accurate quantification of methanogens can inform ecological and clinical insights. By prioritizing primer specificity, researchers can mitigate the biases inherent in universal assays and better elucidate the roles of archaea in microbial ecosystems.

In addition to the selection of highly archaea-specific primers for hypervariable regions of the 16S rRNA gene, the use of primers annealing to functional genes involved in methane metabolism, such as *mcrA* and/or *mtaB1*, appears to be a reasonable option to investigate diversity of methanogens in the human gut [61, 62]. However, the paucity of high-quality, curated sequence databases of such genes complicates the analysis of diversity and ecology from *mcrA*/*mtaB1* profiling data alone. Furthermore, the analysis of exclusively functional genes of methanogenesis fails to provide insight into the overall structure of the microbial community and, consequently, the ecological niche of methanogens within it. Consequently, the analysis of functional genes serves as an additional tool, rather than a substitute, for 16S-profiling.

2.3 Abundance bias due to rRNA operon copy number variation

A significant factor that exerts a substantial influence on the outcomes of quantitative analyses of microbial communities employing 16S profiling is the variation in the copy number of rRNA operons present within the genomes of various taxa of prokaryotes. Given that the number of copies ranges from one to six or seven, taxa that carry a high number of operons appear to be more prevalent than taxa with one or two operons [63]. However, this does not accurately reflect the actual ratio of their cells within the microbiome [42]. A number of software packages, including

PICRUSt2 [64] and Tax4Fun [65], facilitate the incorporation of this factor and perform the normalization of results by the mean number of operons per taxon. However, it should be noted that researchers do not consistently undertake such additional analyses. To address the question of whether variation in rRNA operon copy number significantly influences the detected abundances of methanogenic archaea in gut microbiomes, we performed an additional analysis of a previously published dataset of microbiomes from high methane-producing rats [66]. As shown in **Figure 4**, the normalization of abundances by the average number of rRNA operons per phylum (according to the RiboGrove database) leads to an almost twofold increase of detectable abundances of the Methanomicrobiota, as well as other phyla with a low number of rRNA operons (Elusimicrobiota, Planctomycetota, etc.). At the same time, the abundance of phyla with a high number of rRNA operons (Proteobacteria and Firmicutes) decreased under normalization.

These results underscore that copy number–related biases may play a more crucial role in accurately representing methanogen abundance in the GIT microbiomes.

2.4 Biological reasons for the lack of correlation between emitted methane and methanogen abundance

In addition to the potential presence of technical bias, a number of biological factors may contribute to the absence of a complete correlation between the quantity of methane produced and the abundance of methanogenic archaea within the gut microbiota. In addition to potential technical biases, there are several biological explanations for the absence of a complete correlation between methane generation and the abundance of methanogenic archaea in the gut microbiota. One plausible

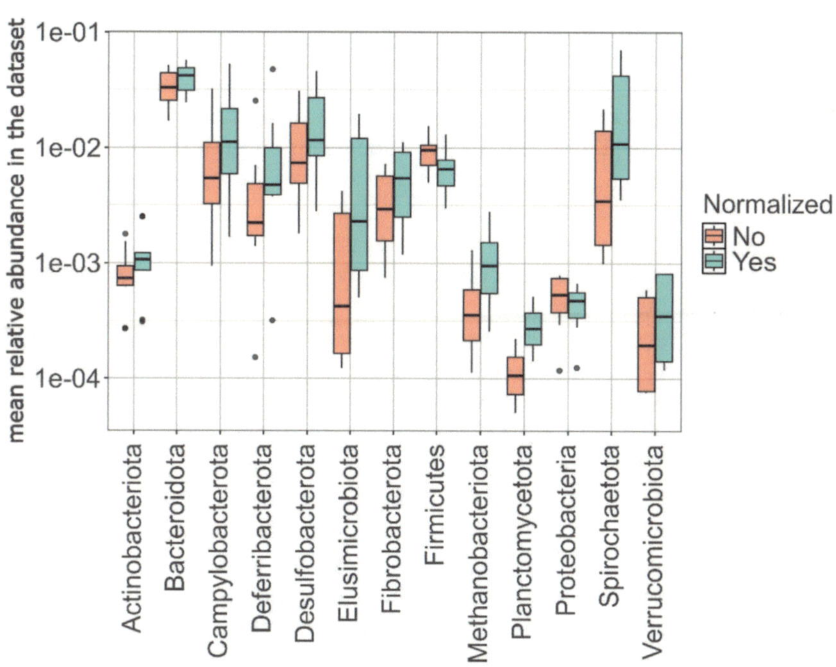

Figure 4.
The relative abundance of different phyla in the dataset from high methane-producing rats with or without normalization.

explanation is the variation in the viability and activity of different taxa within the gut microbial community. For instance, Ben-Amor et al. employed SYTO BC staining and sorting, followed by sequencing of distinct cell fractions, demonstrating that merely 50% of cells in the gut microbiota are active [67]. Later, D'Auria and colleagues exhibited a markedly divergent composition of the active microbiota using pyronin-Y, a fluorescent stain for total RNA, followed by cell sorting and sequencing of the active (stained) fraction compared to the total fraction [68]. The challenges associated with conducting experiments involving methanogenic archaea stem from the absence of adequate staining techniques, thereby hindering the capacity for qualitative sorting of the active archaeal cells. One potential approach involves the analysis of metagenomic RNA, as opposed to rDNA, as implemented in conventional 16S metabarcoding protocols. To the best of our knowledge, however, no studies have hitherto explored this method.

Reliable quantification of methanogenic archaea in animal and human microbiota samples appears to be a cornerstone in understanding their ecological role and impact on health and fitness. However, because they belong to a very distinct group of microorganisms, difficult to culture and poorly represented in animal microbial consortia, the most common methods of microbiota analysis often underestimate their abundance. These difficulties emerge at nearly every stage of sample preparation, including the isolation of metagenomic DNA (methanogenic cells are more difficult to lysate than bacterial cells), the preparation of amplicon libraries (universal primers lack sufficient specificity with respect to methanogenic archaea), and bioinformatic analysis (lack of normalization for the number of rRNA copies of operons). Consequently, these biases overlap and result in frequently elevated levels of methane emissions that do not correspond to the proportion of methanogenic archaea within the gastrointestinal tract microbiota. The most reliable approach to address this issue is to utilize control preparations of synthetic microbial communities containing methanogenic archaea cells (to regulate methane emission) or cloned copies of methanogenic archaea ribosomal RNA genes (to control amplification and bioinformatic analysis) and optimize protocols to ensure the attainment of reliable results. At the same time, biological reasons for the underestimation of the role of archaea in GIT communities cannot be excluded. First, the archaea fraction may well contain a higher proportion of viable cells compared to the bacterial part of intestinal microbial consortia. Second, cells of methanogenic archaea may have higher transcriptional and metabolic activity compared to bacterial cells. However, these assumptions require detailed experimental verification.

According to the results of clinical studies, several main factors can be identified that influence methane production: geographical features, socioeconomic status, age, and gender [69]. In many studies, female gender was also associated with a positive CH4 producer status, but not everywhere. According to the results of published studies, the influence of such a factor as gender on the results of the breath test is contradictory. There are studies confirming the fact that women are more likely to produce methane [70], while other studies did not find any differences in breath test results between men and women [71–74]. At the same time, the study by Newberry et al. [75], noted that with age, women are more likely to develop small intestinal bacterial overgrowth (SIBO) and, as a result, an increase in methane production is determined compared to men. A study by other authors [76] did not reveal any effect of age on the breath test results in women.

3. Factors influencing breath CH_4 production

It is commonly believed that the microbial community in the digestive tract develops during childhood and increases throughout life. Methane is not detected in exhaled air in children under 3 years of age; its detection becomes evident by adolescence, reaching levels similar to those of adults [77]. The clinical studies confirmed that small children, older children, and adults have significant differences in the percentage of CH_4-producing breath [69]. Several studies have indicated that a sharp increase occurs between 10 and 15 years of age, which may indicate a transition from childhood to adolescence [78, 79]. According to Polag, there is a continuous increase in the number of breath CH4 producers with age [79]. The study included 428 people aged 4 to 95 years. It was observed that the percentage of methane in exhaled air increases with age: from 5% in children (1–15 years old) to 57% in the elderly (over 75 years old). When divided into 5-year age intervals, there was a decrease in the percentage of people who emit methane in the sixth and seventh decades. It was found that in the first half of life, the main methane producers are women, who are inferior to men between the ages of 20 and 65. The authors conclude that there is a possible relationship between methanogenic activity and hormonal factors.

One of the first studies to confirm the effect of age on methanogens was conducted on animals (rats) [80]. The authors suggest that one possible explanation for the increase in methanogens with age is a gradual shift of the use of electrons from the reduction of CO2 to acetate by acetogens to the reduction of CO_2 to CH_4. Age-related changes in the composition of archaea and methane production were identified in dairy cows and heifers using high-throughput sequencing [81]. A significant shift in bacterial taxa was observed in relation to age-related changes in the rumen microbiota, but archaeal taxa remained relatively stable. *Prevotella, Ruminococcus, Flavonifractor, Succinivibrio*, and *Methanobrevibacter* were affected by age. Methane production and volatile fatty acid concentrations were age-related. Heifers (9–10 months) had lower methane production but higher methane production per dry matter intake (DMI). The acetate/propionate ratio showed a significant decrease with age. This study revealed different associations between predominant bacterial phylotypes and *Methanobrevibacter* with increasing age. *Prevotella* was strongly correlated with *Methanobrevibacter* in heifers. In cows that are 96–120 months old, this connection was replaced by a correlation between *Succinivibrio* and *Methanobrevibacter*. The age-specific variation in rumen fermentation and methane production may be attributed to this shift.

Age-related changes in methanogen diversity were identified [82]. It was revealed that a new phylogenetic type Mx-01 (*Methanomassiliicoccales*), a new phylotype hypothesized to belong to a putative sixth methanogenic order, is most often observed in the feces of elderly people. The presence of the mcrA gene was noted in only 1 out of 23 newborns. It has been established that the diversity of the archaea composition changes with age, in particular, this applies to the species *Methanobrevibacter smithii*, which was recently divided into two subspecies *M. smithii* and *Ca M. Intestine* [83]. An increase in the representation of methanogens with age with a decrease in diversity was revealed. In the young adults age group, there was an increase in the amount of methanogens mainly due to Ca M. Intestine. A high abundance of *M. smithii* was observed in the older age groups. Age-related "swings" of composition changes of archaea producing butyrate were noted: a decrease in *Lachnospiraceae* with an increase in *Oscillospiraceae* was noted.

Age-related changes in the composition of archaea are also observed in other localizations like the skin [84]. The highest representation of archaea was found in the elderly and people under 12 years of age, compared with middle-aged people. The study authors claim that this fact is closely associated with the skin's lipid composition, pH level, and moisture changes.

Breastfeeding and nutrition have an impact on the formation of the intestinal archaeome during the first years of life. A recently published study on the effect of breastfeeding on the intestinal microbiome of newborns revealed similar beta diversity and metabolic prediction between samples of human colostrum and neonatal stools [85]. This fact suggests the possibility of vertical transmission of microbiota, in particular archaea, from mother to child through breastfeeding, which does not exclude the influence of other factors on the formation of the microbiota of the newborn. An earlier study assessed the effect of dairy products (milk, yogurt, and cheese) on the density of methanogens in children aged 2 to 10 years [86]. It was found that children who regularly consume organic dairy products are five times more likely to be colonized by *M. Smithii* for milk and four times more likely to be colonized for yogurt. The authors identified milk products as possible source for *M. smithii*, but not *M. stadtmanae*.

Based on the results of the above studies, it can be concluded that the microbiota is constantly changing throughout life. Concomitant diseases such as irritable bowel syndrome, often accompanied by constipation and bacterial overgrowth syndrome, may be additional factors determining age-related changes in the microbiota leading to changes in methane production. The increase in the frequency of lactose intolerance observed with age is accompanied by an increase in the production of H_2, which is a substrate for methanogens in the intestine. The age-related increases in concomitant diseases requiring medications that alter secretory activity and intestinal tone are additional factors in methane production change.

4. Archaea and pathology

Over the past decades, the importance of such microorganisms as archaea in the pathogenesis of various diseases has been actively discussed in scientific literature. Until recently, it was believed that archaea are non-pathogenic, that is, they cannot cause the development of a pathological inflammatory reaction of the body, which was possibly due to technical difficulties in identifying these microorganisms. In recent years, sufficient experimental data have been accumulated, thanks to which it can be said that archaea, if not directly, then indirectly, affect the pathogenesis of various diseases, entering into symbiotic relationships with other microorganisms. Any pathological process in a living organism leads to dysbiosis, a change in the composition of metabolites of the microbiota of a living organism. It is quite possible that the observed change in the composition of archaea in various diseases is a reflection of the reaction of the archaea to a change in the symbiotic consortium of archaea with other representatives of the microbiota.

4.1 Archaea and digestive system

According to the literature, the largest number of studies on the involvement in the pathogenesis of various diseases concerns the digestive system. These changes are represented by quantitative and qualitative changes in the composition of archaea. *M. oralis* and *M. smithii* levels were elevated in endodontic infections, whereas M. oralis was elevated in periodontitis [87, 88].

The large intestine pathology confirms a change in methanogen activity, both upwards and downwards, as confirmed by breath test and sequencing. The most significant changes are associated with *Methanobrevibacter*.

In diseases such as irritable bowel syndrome (IBS) with constipation, encopresis, and diverticulosis, an increase in the level of *M. smithii* is observed [3, 89–91]. According to the results of most clinical studies, increased methane production is most often observed in IBS accompanied by constipation [92, 93]. However, in a study including 87 children aged 13 ± 2.6 years, of whom 50 (57.5%) were female, no relationship was found between methane and hydrogen production with the IBS subtype, or the frequency/severity of abdominal pain, or psychosocial distress [94]. Methane production was positively correlated with the transit time through the entire intestine ($r = 0.31$, $P < 0.005$) and had a negative relationship with the frequency of bowel movements ($r = -0.245$, $P < 0.05$). This fact may serve as a basis for recognizing the total methane level in the lactulose breath test as a biomarker of the total intestinal transit time and frequency of bowel movements in children with IBS. At the same time, in the study by Pimental, increased methane production was noted in adult patients with IBS, especially when IBS was combined with constipation [92]. Oral administration of neomycin reduced methane production and clinical manifestations of constipation. In another study [95], no relationship was found between the severity of constipation and the methane level. A weak correlation was found between total gas levels and several IBS symptoms: bloating ($r = 0.324$, $P = 0.039$), flatulence ($r = 0.314$, $P = 0.046$), and abdominal pain ($r = 0.364$, $P = 0.018$).

Methane-predominant SIBO is associated with an increased risk of developing hepatocellular carcinoma in patients with liver cirrhosis and related complications, including hepatic encephalopathy [96].

According to published data, patients with appendicular abscess have an increase in *M. oralis and M. smithii* [97]. After appendectomy, the level of methanogens decreased, which was confirmed by a breath test [98]. There is a group of diseases in which the level of methanogenesis is reduced. These are inflammatory bowel diseases, such as non-specific ulcerative colitis (UC) [99] and Crohn's disease, and diseases associated with a decreased transit time of food through the intestine, such as IBS with diarrhea [88]. In 2021, a meta-analysis of 17 clinical studies including 1653 patients with SIBO and 7 clinical studies including 626 patients with inflammatory bowel diseases (IBD) (Crohn's disease and ulcerative colitis (UC)) was published [93]. In all studies, a lactulose breath test was performed to assess bacterial overgrowth syndrome. The analysis revealed a statistically significant increase in methane levels in patients with IBS compared to patients with IBD. The absence of methane production was found in patients with Crohn's disease compared to UC. Methane levels were correlated with the severity of constipation in patients with IBS.

Severe malnutrition has been accompanied by a decrease in the level of archaea and a reduction in bacterial composition due to reduced nutrient supply for microorganisms [100]. With anorexia nervosa, an increase in the level of *M. smithii* is observed [101].

4.2 Archaea and cardiovascular system

4.2.1 Archaea, atherosclerosis, and TMAO

It has been established that atherosclerosis is a chronic arterial disease manifested by the appearance of atherosclerotic plaques, which, as the disease progresses, can lead to adverse cardiovascular outcomes (heart attack, stroke, and gangrene of the lower extremities). There are several risk factors for the development of atherosclerosis, among which diet occupies an important place. In recent years, more and more attention has been paid to the participation of microbiota in maintaining the normal functioning of the cardiovascular system [102–105]. The formation of intestinal microbiota occurs, among other things, under the influence of diet. Increased consumption of foods rich in choline and L-carnitine leads to the formation of the metabolite trimethylamine (TMA) in the intestine, which is absorbed into the blood and enters the liver. In the liver, under the action of the enzyme flavin monooxygenase (FMO), TMA is converted into trimethylamine-N-oxide (TMAO), which has a proatherogenic effect [32, 106–108]. In individuals with a hereditary defect in flavin-containing monooxygenase 3, bacterial TMA production is believed to contribute to the symptoms of trimethylaminuria (TMAU, fish-odor syndrome) [109]. The formation of TMAO in plasma (pTMAO) originates from intestinal TMA, which is created by the interaction of diet, gut microbiota, and the human host. Atherosclerosis risk may be affected by TMA metabolism by altering trimethylamine oxide (TMAO) production levels. The literature discusses two main ways to reduce the formation of TMAO, which affect the metabolism and absorption of TMA at the intestinal level [110]. In 2014, the concept of "archaeobiotics" was introduced—a new generation of symbiotics, in particular methanogenic archaea, or prebiotics that stimulate the development of methanogenic archaea in the intestine [24]. It has been demonstrated that natural methanogenic archaea from the human gut can reduce TMA with hydrogen for methanogenesis. This process was described over 40 years ago in the rumen of cows [111]. However, delivering sufficient amounts of these oxygen-sensitive microorganisms to the gut remains an important hurdle. The technique was patented, but so far it has not been tested in humans [30].

A study by Brugère et al. experimentally confirmed that *Methanomassiliicoccus luminyensis B10* was able to reduce TMA with H_2 for methanogenesis. The authors propose that metabolic disorders could benefit from the use of this archaea strains as treatments [31]. Another way to decrease TMA involves engineering probiotic strains to metabolize it, or live engineered biotherapeutic products [110].

Ramezani et al. investigated the efficacy of gut colonization with methanogenic archaea (MA) on lowering plasma TMAO concentrations [31]. It was discovered that five MA species have the ability to colonize and love plasma TMAO levels in C57BL/6 mice on high choline/TMA supplemented diet. Gut colonization with *M. smithii* results in a significant decrease of plasma TMAO levels, with a trend to reduce atherosclerosis burden in Apoe−/−mice as 44% reduction of aortic plate area and 52% reduction of fat content in atherosclerotic plaques.

Borrel G., et al. show that *Methanomassiliicoccales* are present in published microbiome datasets from eight countries. Four of the six new *Methanomassiliicoccales* genomes assembled from ELDERMET metagenomes contained genes for utilizing trimethylamine (TMA). The abundance of TMA-utilizing *Methanomassiliicoccales* was positively correlated with the count of bacterial genes for TMA production and negatively with fecal TMA concentrations [112]. It was found that the presence of

Methanomassiliicoccales in human populations was not universal, but they were associated with bacteria-produced trimethylamine [113].

4.2.2 Archaea and cardiovascular diseases

A recently published study assessed the prevalence and identified risk factors for IBS in patients with acute ischemic stroke (first 48 hours) [114]. A positive SIBO test result was recorded in 23 of the 80 patients (28%) diagnosed with acute ischemic stroke. According to statistical analysis (multivariate logistic regression analysis), triglyceride (TG) and homocysteine (Hcy) levels were identified as independent risk factors for the development of SIBO in patients with acute ischemic stroke ($P < 0.005$), with high predictive value in patients with acute ischemic stroke, especially when both TG and Hcy were taken into account simultaneously.

A large population-based study by Takakura et al. assessed the effect of methane on resting heart rate [115]. The observation period was 14 weeks. The study included 1125 subjects. The study confirmed the theory of the effect of methane on the cholinergic system: a relationship was found between a decrease in the level of methane in exhaled air during 14 weeks of observation and an increase in heart rate.

During a prospective study by Mollar et al. the relationship between the content of hydrogen (H_2) and methane (CH_4) in exhaled air as a diagnostic criterion for SIBO and manifestations of heart failure was assessed [116]. The study involved 102 patients with a breath test with lactulose. Gas accumulation was quantified by the area under the curve of CH_4 (AUC-CH_4) and H_2 (AUC-H_2). Clinical endpoints included the composite of all-cause death with all-cause or HF hospitalizations, recurrent all-cause hospitalizations, and recurrent HF hospitalizations. In the analysis, there was an association between AUC-H2 (per 1000 U) and all-cause death/all-cause hospitalization (hazard ratio [HR] 1.21, 95% CI 1.04–1.40; $P = 0.012$), all-cause death/HF hospitalization (HR 1.20, 95% CI 1.03–1.40; $P = 0.021$), and an increase in all-cause readmissions (IRR 1.31, 95% CI 1.14–1.51; $P < 0.001$) and HF-related readmissions (IRR 1.41, 95% CI 1.15–1.72; $P = 0.001$). AUC-CH4 was not associated with any of these endpoints. The study concluded that exhaled hydrogen breath testing is a safe and non-invasive method for assessing SIBO and a marker of higher risk of long-term adverse clinical events in patients with HF, in contrast to AUC-CH4, which did not show any prognostic value.

However, 2 years later, the results of another study were published, which confirmed the prognostic value of such an indicator as methane production in terms of adverse outcomes in HF [117]. The study included 287 patients hospitalized for heart failure at Fudan University Hospital. Patients were recruited between 2017 and 2019. SIBO was diagnosed in 128 patients. At the initial assessment, the SIBO group of patients had clinical manifestations of HF according to the NYHA III-IV scale. In patients with positive SIBO and HF with reduced ejection fraction, the risk of rehospitalization for HF increased by 2.77 times (OR 2.77; 95% CI 1.62–4.74; $P < 0.001$) without significant differences in cardiovascular mortality (OR 1.66; 95% CI 0.40–6.94; $P = 0.467$). In patients with positive SIBO in HF with preserved ejection fraction, the risk of cardiovascular mortality was increased (OR 7.34; 95% CI 1.58–34.13; $P = 0.011$), without a significant effect on the frequency of rehospitalization for HF (OR 3.03; 95% CI 0.98–9.38; $P = 0.077$). Based on the study results, the authors conclude that SIBO is an independent risk factor for assessing the primary endpoint in patients with HF (odds ratio [OR] 2.13; 95% CI; 1.26–3.58; $P = 0.005$). It was the elevated methane level in the context of SIBO that demonstrated prognostic value for adverse outcomes in HF (OR 2.35; 95% CI 1.38–4.02; $P < 0.001$), while no similar relationship was found between SIBO with increased hydrogen production and outcomes.

In a recently published study, we estimate the influence of enzymatic activity of microbiota on hemodynamics, arterial stiffness, and emotional state of patients with mild-to-moderate risk of cardiovascular complications [118]. The study included 22 patients with mild-to-moderate risk of cardiovascular complications. In addition to standard examination, patients underwent volumetric sphygmography with analysis of cardio-ankle vascular index, HADS scores, and results of hydrogen-methane breath test with lactulose. It was revealed that there was a negative correlation between anxiety and H2 ($r = -0.44$; $p = 0.05$), as well as H_2/CH_4 ratio ($r = -0.53$; $p = 0.02$). H_2 level was a negative predictor of anxiety independent of age, systolic and diastolic blood pressure, as well as heart rate ($\beta = -0.64$; $p < 0.01$). There was a negative correlation between CAVI and exhaled H_2 ($r = -0.41$; $p < 0.05$). CH4 AUC was an age-independent predictor of higher systolic blood pressure ($\beta = -0.49$; $p = 0.03$).

4.3 Archaea and endocrine system

In recent years, the issue of the participation of intestinal microbiota in the pathogenesis of many somatic and endocrine diseases has been actively discussed. In diabetes mellitus type II, intestinal dysbiosis with a decrease in saccharolytic bacteria is observed. It is not surprising that the consequence of this is a change in the composition of archaea in the form of an increase in the composition of methanogens [119]. In diabetes mellitus type I, there was an association between elevated levels of methanogens and higher serum glucose levels [120].

The syndrome of excessive bacterial growth accompanies many diseases, including endocrine pathology. A recently published study evaluated the efficacy of probiotics and prebiotics in treating SIBO associated with subclinical hypothyroidism (SCH) in the second trimester of pregnancy [121]. The study included 78 pregnant women with thyroid pathology and 74 healthy pregnant women in the second trimester of pregnancy. Patients were analyzed for high-sensitivity C-reactive protein (hsCRP) levels, lipid metabolism parameters, and lactulose methane–hydrogen breath test, and the severity of gastrointestinal symptoms was assessed using the GSRS scale. In the hypothyroidism group, 32 patients with SIBO were selected as the intervention group. The course of probiotics + prebiotics therapy was 21 days. The parameters were assessed before and after therapy. Initially, patients with hypothyroidism had a high incidence of SIBO with clinical manifestations of dyspepsia and constipation, high production of hydrogen and methane, as well as the elevated level of highly sensitive C-reactive protein and a total score on the GSRS scale ($p < 0.05$). After treatment, serum levels of thyroid stimulating hormone (TSH), total cholesterol (TC), triglycerides (TG), low-density lipoproteins (LDL), and hsCRP in the group of women receiving prebiotics + probiotics were reduced, and high-density lipoproteins (HDL) were increased compared to pre-treatment values ($P < 0.05$). Therapy with prebiotics and probiotics led to a decrease in methane and hydrogen production, normalization of clinical manifestations in the form of a decrease in dyspepsia, constipation, and a total score on the GSRS scale ($p < 0.05$).

In obesity, there are inconsistent changes in methanogen levels, as reported in the literature by sequencing and breath testing [100, 122–124]. Increased diversity of *Methanobacteriales* has been reported in obese subjects compared to normal weight.

Following weight-reducing surgeries, *Methanobacteriales* density has been reported to be reduced [125]. One hypothesis for this condition is that interspecies H_2 transfer between bacteria and archaea leads to increased energy absorption by the colon in obese subjects [100]. A study by Mathur et al. assessed the relationship between methane and hydrogen in a breath test, body weight, and body fat percentage in the general population [126]. The study included 792 patients divided into 4 groups based on hydrogen and methane production in exhaled air: normal (N) (methane <3 ppm and hydrogen <20 ppm at 90 minutes or earlier); hydrogen positive only (H+) [methane <3 ppm and hydrogen ≥20 ppm]; methane only (M+) positivity (methane ≥3 ppm and hydrogen <20 ppm); or methane and hydrogen positivity (M+/H+) (methane ≥3 ppm and hydrogen ≥20 ppm]. There were significant differences in age but not in gender between the groups. After accounting for age as a confounding variable, M+/H+ subjects had a significantly higher BMI than the other groups ($P < .02$) and also had a significantly higher percentage of body fat ($P < .001$).

According to the literature, the effect of weight on breath test results is contradictory. In studies on adult patients, a correlation was found between increased weight and a positive breath test [123]. However, in other studies, Cortez APB., et al., [127] and Santos ANDR, et al. [71], noted the opposite trend in children: with a positive breath test as confirmation that SIBO was associated with weight loss. In the studies of Kiow et al. [76], no effect of BMI on the results of the breath test was found, and in the study of Kim [70], a weak relationship was found between BMI and a positive breath test in women. An interesting study was conducted in Spain in 2021, which assessed the relationship between intestinal microbiota, including the representation of *M. smithii* and metabolic disorders in patients with and without pets [128]. Patients with pets were less likely to be obese. As for the intestinal microbiota, they had an increased density of *Methanobrevibacter, Coprococcus, and Oscillospira*, and among the bacteria, *Serratia* was most often found. People without pets had an increase in *Ruminococcus, Enterobacteriaceae*, and *Anaerotruncus*. This group was more likely to have an increase in body weight.

In our study, patients with type 2 diabetes mellitus and obesity showed a decrease in hydrogen levels according to the breath test, as well as a positive correlation between the level of postprandial glycemia and methane levels. In patients in the control group, an inverse correlation was observed between the level of glycemia after taking dietary fiber (guar gum) and the level of hydrogen in exhaled air, which was not observed in patients with type 2 diabetes mellitus. These changes may be associated with a decrease in the level of hydrogen-producing bacteria and an increase in the activity of methanogens [129].

4.4 Archaea and neurodegenerative diseases

In neurodegenerative diseases, changes in the intestinal microbiota are observed [130, 131]. Parkinson's disease is accompanied by decreased intestinal motility, constipation, and changes in the intestinal microbiota in the form of an increased representation of methanogens [132].

The latest published study addresses the issue of the influence of archaea on human cognitive functions [133]. Using shotgun metagenomics and neurophysiological test, the relationship between the microbial community and cognitive functions was established in the studied cohort (IRONMET, n = 125). *M. smithii* and cognitive functions were identified. Depending on the representation of *M. smithii*, a change in the bacterial composition of the microbiota was observed. Thus, representatives with a high level of *M. smithii* are characterized by high levels of *Verrucomicrobia, Synergistetes, and Lentisphaerae* and a low level of *Bacteroides* and *Proteobacteria*. Differences in microbial composition determined differences in metabolites. Thus, in the participants with high abundance of *M. smithii*, enrichment in energy, butyrate, and bile acid metabolism was observed. The blood plasma levels of methylhistidine, phenylacetate, alpha-linolenic and linolenic acids were increased, as well as the levels of 3-methylhistidine, phenylacetylglutamine, adrenic acid and isolithocholic acid.

4.5 Archaea and cancer

Conflicting data concern the participation of methanogens in the pathogenesis of colorectal cancer [134]. Early studies found that patients with high levels of methanogenesis according to a breath test had a lower risk of developing colorectal cancer [134]. Later, it was found that there was a positive correlation between the presence of halophilic bacteria, prostate adenoma, and colorectal cancer [135]. There was an increase in the representation of halophilic bacteria with a decrease in methanogens in the comparative series: control group, patients with adenoma, and patients with colorectal cancer. It was established that *Thaumarchaeota* is involved in the biochemical processes of oxidation of ammonia, a product of protein metabolism in the process of carcinogenesis [136]. In a large-scale study, when analyzing more than 44,000 contigs across different body sites, a relationship was established between human-associated archaea and cancer-associated metabolites [137]. The greatest diversity of cancer-associated metabolites was found in the oral cavity and intestine.

4.6 Archaea and the immune system

Until recently, the question of immunogenicity in archaea was not entirely clear given the characteristics of structure and absence of structures characteristic for bacterial cells that determine the immune response, namely lipopolysaccharides, lipoproteins, and peptidoglycan. Two receptors have now been identified on human immune cells for archaea identification: TLR8 [138] and MINCLE [139]. The largest number of publications is about methanogens that confirm an immune system response to archaea through a humoral and cellular response [61, 140]. The response of humoral immunity has been confirmed in experimental studies on animals (mice), as well as in clinical trials on patients with inflammatory bowel diseases and periodontitis. For the first time, an increase in the level of IgG to *M. oralis* was noted in patients with periodontitis [87]. Later, the same authors published results confirming the cross-immune reaction between antigenic group II chaperonin in *M. oralis* and human chaperonin CCT [141]. An increase in the level of IgG to *M. stadtmanae* with a simultaneous more frequent detection of MSS in feces was observed in patients with inflammatory bowel diseases (IBD) [142]. When stimulating peripheral blood mononuclear cells with *M. stadtmanae*, an increase in TNF-α secretion was noted.

An increase in IgG levels was observed in rodents (mice) in the model of bioaerosol-related pulmonary diseases [143]. The animals were administered intranasally a suspension of *M. smithii* or *M. stadtmanae* cultures in different doses (6.25, 25, or 100 mg) once a day for 3 days over 3 weeks. During the study, it was established that the type of archaea determines the strength and nature of the immune response in the lungs. However, both types of archaea caused similar titers of antigen-specific IgG in plasma. In other respects, *M. stadtmanae* was more pathogenic, as it caused more serious histopathological changes than *M. smithii*.

In further studies by these authors on this model of pathology of the bronchopulmonary system, the pro-inflammatory effect of *M. smithii* on the state of the respiratory tract, accompanied by activation of Th17, was revealed. This type of reaction corresponds to a type IV hypersensitivity reaction that does not require the participation of eosinophils. The latter was confirmed in experiments on mice with eosinophil and mast cell deficiency [144].

German researchers studied the interaction of archaea and antimicrobial peptides (AMP) [145, 146]. The effects of different synthetic AMPs on the growth of *M. smithii*, *M. stadtmanae*, and *Methanosarcina mazei* were tested. All three tested methanoarchaea were highly sensitive against derivatives of porcine NK-lysin (NK2 and C7S) and a synthetic anti-lipopolysaccharide (anti-LPS) peptide, Lpep 19–2.5. The potent AMP concentrations affecting growth were below 10 μM. The highest concentrations of AMP were required for the destruction of MSS. Within 4 hours of incubation with AMPs, the structural integrity of methanoarchaea cells was destroyed, as revealed by atomic force microscopy and transmission electron microscopy. Their next step was to investigate the immunogenic qualities of *Methanomassiliicoccus luminyensis* and its susceptibility to antimicrobial peptides [146]. Because of its remarkable metabolic properties, in particular the degradation of trimethylamines, this strain was intended to be used as an 'archaeobiotic' for human intestinal metabolic disorders. It was demonstrated that *M. luminyensis* was highly sensitive against LL32, a derivative of human cathelicidin. The data obtained confirm that *M. luminyensis* is a typical commensal gut microbe.

In recent years, the results of studies on cell cultures (monocyte-derivate dendrite cells (MoDCs) and peripheral blood mononuclear cells (PBNCs)) have been published; the main purpose of which was to clarify the possible microbe-associated molecular patterns (MAMPs) and the pattern recognition receptors (PRRs), characteristic of archaea.

Bang et al. studied immune reaction of intestinal epithelial cells (intestinal epithelial cell line Caco-2/BBe) in response to *M. stadtmanae* and *M. smithii* stimulation [145]. Stimulation with *M. stadtmanae* or *M. smithii* did not result in the release of cytokine IL-8 or significant changes in transcript levels of genes encoding TNF-a, IL-8, human beta defensin 1 (HBD1), HBD4, human defensin 6 (HD6), or human cathelicidin LL37. It can be concluded from these findings that *M. stadtmanae* and *M. smithii* are not recognized by human intestinal epithelial cells. Activation of MoDCs in response to *M. stadtmanae* leads to a significant release of cytokines TNF-α and IL-1β, while the reaction to *M. smithii* is weak. It was revealed that *M. stadtmanae* and *M. smithii* increased expression of CD 86 and CD 197 receptors with a more pronounced response to *M. stadtmanae*. *M. stadtmanae* altered the expression of antimicrobial peptide genes in MoDC to varying degrees.

Subsequent studies by this group of scientists confirmed the data that the interaction of archaea and cells of the immune system is similar to the antiviral immune response, which is carried out through phagocytosis with the formation of a lysosome followed by the expression of interferon genes: IFN-α14, INF-β, and INF-γ1 [138]. *M. stadtmanae* induced an antiviral type I/III IFN response. It turned out that it is RNA that has immunogenicity, stimulating the secretion of TNF-α and IL-1β in PBMCs and MoDC cells. TLR8 engagement is the sole factor that activates the NLRP3 inflammasome, even though *M. stadtmanae* recognition is mediated by TLR7 and TLR8. This process is similar to the hallmarks of both the canonical and the recently discovered alternative inflammasome activation (pyroptosis).

The results of recent publications have shown that archaeal lipids were recognized by the C-type lectin receptor MINCLE and induced immune response [139].

5. Strategies for decreasing CH_4 emissions

According to the literature, an increase in methanogens, accompanied by an increase in the level of methane in exhaled air, is recognized as one of the predictors of the development of SIBO, which worsens the course of many diseases, including cardiovascular diseases, endocrine pathology, and neurodegenerative diseases. All of the listed changes confirm the fact of the interaction of the intestinal microbiota with the organs and systems of the body, determining the functioning of the axes: "gut-brain," "gut-lungs," "gut-cardiovascular system," and "gut-endocrine system." A decrease in methanogenesis with an increase in the level of hydrogen as an antioxidant is one of the points of influence on these axes and the regulation of the function of the cardiovascular system, endocrine system, central nervous system, and bronchopulmonary system. Currently, several ways of influencing methanogenesis in humans are considered in the literature:

1. Rifaximin and neomycin administration [6]

2. Statin administration: lovastatin and mevastatin (**Figure 5**) [147, 148]

3. A low FODMAP diet [13, 16, 149]

4. Probiotic administration: capsules containing the strain *Lactobacillus reuteri* [150]

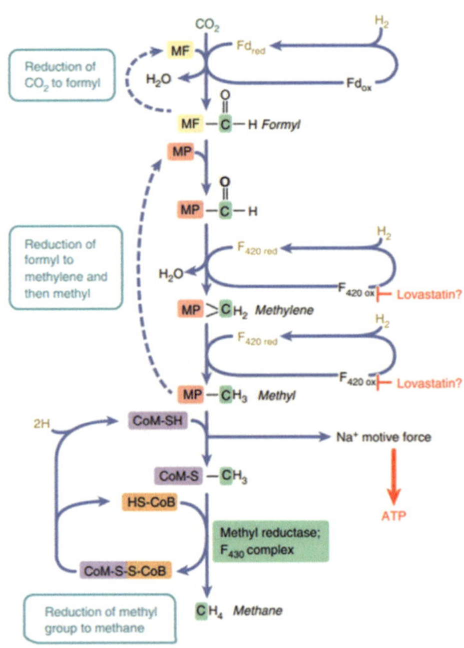

Figure 5.
Methanogenesis pathway and potential role of lovastatin. Lovastatin and mevastatin had higher affinities for the F420 binding site on fno than did F420 itself. Lovastatin may act as inhibitor of F420-dependent NADP oxidoreductase (fno), resulting in less methane formation. Reduced carbon atom is highlighted in green. MF, methanofuran; MP, methanopterin; CoM, coenzyme M; Fd, ferredoxin; CoB, coenzyme B. This figure was given from Ref. [147].

Currently, it is believed that 30% of anthropogenic methane is formed in the intestines of animals, among which the leaders are ruminant livestock [151]. Methane is formed during the metabolism of methanogenic archaea, which use hydrogen and other products of microbial fermentation in the intestine. This process helps remove waste products of intestinal microorganisms that limit the growth of microflora. Methanogenesis results in a loss of 6–10% of energy consumed by ruminants raised worldwide. Methanogenesis is undesirable from the standpoint of energy efficiency and feed costs [152]. CH_4 emissions from ruminant livestock can be decreased through strategies such as animal management, genetic selection, rumen microbiome manipulation, and nutritional modulation.

Ways to reduce methanogenesis in farm animals (primarily ruminant livestock) [61, 153].

1. Diet modification:

   - Addition of various dietary supplements (rapeseed oil, sweet potato vine silage, and alfalfa). In this case, the time of food residence in the rumen has an effect: a decrease in time reduces methanogenesis [154, 155].

   - Different diets, dietary modification. High fiber content compared to high starch content leads to an increase in the acetate/propionate ratio and an increase in cellulolytic bacteria levels and an increase in methane levels [156]. Meanwhile, an increase in amilum leads to an increase in the activity of amylolytic bacteria and an increase in rumen H2 levels, and a decrease in methane levels. The other way to reduce CH4 production is to increase dietary fats or fat-rich supplements [155].

2. Adding chemical compounds:

   - Bromochloromethane results in a 30% reduction in methane levels [157].

   - Bromoform, contained in the seaweed Asparaguses species, reduces methanogenesis by 90% [158]. The disadvantage of this approach is the toxicity of bromoform, which manifests itself in the carcinogenesis, as well as ozone-depleting properties. The compound has increased volatility and enters the atmosphere, where it turns into bromine, which destroys ozone. Bromoform can accumulate in animal meat and milk consumed by humans. Despite its high efficiency in reducing methanogenesis, consumption of this compound by cattle leads to animals refusing to eat.

   - 3-nitrooxypropanol, which affects the enzyme methyl-coenzyme M reductase (MCR) [159], leads to a decrease in methane production by 30–82% [160, 161].

   - Ionophores and organic acids. Ionophores usually decrease CH4 in the rumen by decreasing H2 production. But ionophore antibiotics (maduramicin, monensin, lasalocid, and salinomycin; usually used as anticoccidials) stimulated caecal methanogenesis in rabbits due to the H_2-dependent formation of acetate inhibition, leading to an increase in the H_2 available for methanogens [162].

3. Changes in the composition of the intestinal microbiota:

    - An increase of Acetobacter competing with methanogens for hydrogen in the synthesis of acetate instead of methane [163]. This type is found among macropodids in Australia. In the stomach of kangaroos, the amount of acetogens predominates; therefore, the minimum amount of methane production is observed.

    - Defaunation results in a 42–80% reduction in methanogenesis [164]. Protozoa act as hosts for archaea, protecting them and providing them with hydrogen for methane synthesis.

    - Use of prebiotics, for example in non-ruminants (pigs)—*Saccharomyces cerevisiae* YST2 [165].

4. Selection of animals with reduced methanogenesis [166].

5. The latest direction for reducing methanogenesis is vaccination against methanogens [167].

6. Conclusions

In the available literature, much more attention is paid to the mechanisms of hydrogen production by the gut microbiota, its role in SIBO, and its antioxidant, anti-apoptotic, and anti-inflammatory properties. The physiological role of methane and mechanisms of its production and potential role in pathology development are much less evident. It depends on the difficulties in sequence determination of the methanogenic archaea and in the long-term assumption that archaea are not pathogenic for human organism. We demonstrate in our review that there are physiological and metabolic differences between groups of humans with a high baseline CH_4 concentration and low baseline H_2 concentration vs. a low CH_4 producer group with low baseline CH_4 concentration and high baseline H_2 concentration. Humans with high production of methane by the gut microbiota predominantly are characterized by constipation, lower levels of short-chain fatty acids in the gut, lower reactivity to the administration of food fibers, and many others. We assume that measurements of the methane levels in the exhaled air or in the flatus might be more affordable in clinical settings compared with the sequencing for the determination of high methane producers. Concentrations of hydrogen and methane in exhaled air during wildly used breath test need high sensitive sensors due to their low levels in ppm range. Non-invasive measurements of hydrogen and methane in intestinal gas (flatus) seem more preferable because low sensitive, cheap sensors could be used (concentrations in % range). Wide screening of the high methane-producing humans will allow the development of personalized recommendations for healthy nutrition and therapy for constipation and develop new approaches for inhibition of excessive methane production that will be followed by the increase in the level of hydrogen and improvement of anti-oxidative defense.

Conflict of interest

The authors declare no conflict of interest.

Author details

Oleg Medvedev[1,2], Oxana Povarova[1*], Nikita Fadeev[3], Ghizal Fatima[4], Ustin Zolotikov[5] and Stepan Toshchakov[5]

1 Department of Pharmacology, Lomonosov Moscow State University, Moscow, Russia

2 Laboratory of Experimental Pharmacology, National Medical Research Center of Cardiology, Moscow, Russia

3 Department of Medical and Technical Information Technologies, N. E. Bauman Moscow State Technical University (a National Research University), Moscow, Russia

4 Department of Biotechnology and Head Chronobiology Unit, Era University, Lucknow, India

5 Center for Genome Research, National Research Center "Kurchatov Institute", Moscow, Russia

*Address all correspondence to: oxpovarova@gmail.com

References

[1] Kalantar-Zadeh K, Berean KJ, Burgell RE, Muir JG, Gibson PR. Intestinal gases: Influence on gut disorders and the role of dietary manipulations. Nature Reviews. Gastroenterology & Hepatology. 2019;**16**(12):733-747. DOI: 10.1038/s41575-019-0193-z. Epub 2019 Sep 13

[2] Mahnert A, Blohs M, Pausan MR, Moissl-Eichinger C. The human archaeome: Methodological pitfalls and knowledge gaps. Emerging Topics in Life Sciences. 2018;**2**(4):469-482. DOI: 10.1042/ETLS20180037

[3] Hwang L, Low K, Khoshini R, Melmed G, Sahakian A, Makhani M, et al. Evaluating breath methane as a diagnostic test for constipation-predominant IBS. Digestive Diseases and Sciences. 2010;**55**(2):398-403. DOI: 10.1007/s10620-009-0778-4

[4] Rezaie A, Buresi M, Lembo A, Lin H, McCallum R, Rao S, et al. Hydrogen and methane-based breath testing in gastrointestinal disorders: The north American consensus. The American Journal of Gastroenterology. 2017;**112**(5):775-784. DOI: 10.1038/ajg.2017.46

[5] Low K, Hwang L, Hua J, Zhu A, Morales W, Pimentel M. A combination of rifaximin and neomycin is most effective in treating irritable bowel syndrome patients with methane on lactulose breath test. Journal of Clinical Gastroenterology. 2010;**44**:547-550. DOI: 10.1097/MCG.0b013e3181c64c90

[6] Pimentel M, Chang C, Chua KS, Mirocha J, DiBaise J, Rao S, et al. Antibiotic treatment of constipation-predominant irritable bowel syndrome. Digestive Diseases and Sciences. 2014;**59**(6):1278-1285. DOI: 10.1007/s10620-014-3157-8

[7] Teigen L, Mathai PP, Matson M, Lopez S, Kozysa D, Kabage AJ, et al. Methanogen abundance thresholds capable of differentiating in vitro methane production in human stool samples. Digestive Diseases and Sciences. 2021;**66**(11):3822-3830. DOI: 10.1007/s10620-020-06721-5

[8] Di Stefano M, Mengoli C, Bergonzi M, Klersy C, Pagani E, Miceli E, et al. Breath methane excretion is not an accurate marker of colonic methane production in irritable bowel syndrome. The American Journal of Gastroenterology. 2015;**110**(6):891-898. DOI: 10.1038/ajg.2015.47

[9] McKay LF, Eastwood MA, Brydon WG. Methane excretion in man–a study of breath, flatus, and faeces. Gut. 1985;**26**:69-74. DOI: 10.1136/gut.26.1.69

[10] King TS, Elia M, Hunter JO. Abnormal colonic fermentation in irritable bowel syndrome. Lancet. 1998;**352**(9135):1187-1189. DOI: 10.1016/s0140-6736(98)02146-1

[11] Benoit SL, Maier RJ, Sawers RG, Greening C. Molecular hydrogen metabolism: A widespread trait of pathogenic bacteria and protists. Microbiology and Molecular Biology Reviews. 2020;**84**(1):e00092-e00019. DOI: 10.1128/MMBR.00092-19

[12] Espey MG. Role of oxygen gradients in shaping redox relationships between the human intestine and its microbiota. Free Radical Biology & Medicine. 2013;**55**:130-140. DOI: 10.1016/j.freeradbiomed.2012.10.554

[13] Ong DK, Mitchell SB, Barrett JS, Shepherd SJ, Irving PM, Biesiekierski JR, et al. Manipulation of dietary short chain carbohydrates alters the pattern of gas production and genesis of symptoms in irritable bowel syndrome. Journal of Gastroenterology and Hepatology. 2010;**25**(8):1366-1373. DOI: 10.1111/j.1440-1746.2010.06370.x

[14] David LA, Maurice CF, Carmody RN, Gootenberg DB, Button JE, Wolfe BE, et al. Diet rapidly and reproducibly alters the human gut microbiome. Nature. 2014;**505**(7484):559-563. DOI: 10.1038/nature12820

[15] Asnicar F, Berry SE, Valdes AM, Nguyen LH, Piccinno G, Drew DA, et al. Microbiome connections with host metabolism and habitual diet from 1,098 deeply phenotyped individuals. Nature Medicine. 2021;**27**(2):321-332. DOI: 10.1038/s41591-020-01183-8

[16] van Lanen AS, de Bree A, Greyling A. Efficacy of a low-FODMAP diet in adult irritable bowel syndrome: A systematic review and meta-analysis. European Journal of Nutrition. 2021;**60**(6):3505-3522. DOI: 10.1007/s00394-020-02473-0. Epub 2021 Feb 14. Erratum in: Eur J Nutr. 2021;60(6):3523. DOI: 10.1007/s00394-021-02620-1

[17] Rumessen JJ, Gudmandhøyer E. Functional bowel-disease: Malabsorption and abdominal distress after ingestion of fructose, sorbitol, and fructose-sorbitol mixtures. Gastroenterology. 1988;**95**:694-700. DOI: 10.1016/S0016-5085(88)80016-7

[18] Shin W. Medical applications of breath hydrogen measurements. Analytical and Bioanalytical Chemistry. 2014;**406**:3931-3939. DOI: 10.1007/s00216-013-7606-6

[19] Aoki Y. Increased concentrations of breath hydrogen gas in Japanese centenarians. Anti-Aging Medicine. 2013;**10**:101-105

[20] Gottlieb K, Le C, Wacher V, Sliman J, Cruz C, Porter T, et al. Selection of a cut-off for high- and low-methane producers using a spot-methane breath test: Results from a large north American dataset of hydrogen, methane and carbon dioxide measurements in breath. Gastroenterology Report. 2017;**5**(3): 193-199. DOI: 10.1093/gastro/gow048

[21] Modesto A, Cameron NR, Varghese C, Peters N, Stokes B, Phillips A, et al. Meta-analysis of the composition of human intestinal gases. Digestive Diseases and Sciences. 2022;**67**:3842-3859. DOI: 10.1007/s10620-021-07254-1], 10.1007/s10620-021-07254-1]

[22] Stewart JA, Chadwick VS, Murray A. Carriage, quantification, and predominance of methanogens and sulfate-reducing bacteria in faecal samples. Letters in Applied Microbiology. 2006;**43**:58-63. DOI: 10.1111/j.1472-765X.2006.01906.x

[23] Takakura W, Pimentel M, Rao S, Villanueva-Millan MJ, Chang C, Morales W, et al. A single fasting exhaled methane level correlates with fecal methanogen load, clinical symptoms and accurately detects intestinal methanogen overgrowth. The American Journal of Gastroenterology. 2022;**117**(3):470-477. DOI: 10.14309/ajg.0000000000001607

[24] Woese CR, Kandler O, Wheelis ML. Towards a natural system of organisms: Proposal for the domains archaea, bacteria, and Eucarya. Proceedings of the National Academy of Sciences of the United States of America. 1990;**87**:4576-4579. DOI: 10.1073/pnas.87.12.4576

[25] Koskinen K, Pausan MR, Perras AK, Beck M, Bang C, Mora M, et al. First

insights into the diverse human archaeome: Specific detection of archaea in the gastrointestinal tract, lung, and nose and on skin. MBio. 2017;**8**:e00824-e00817. DOI: 10 .1128/ mBio.00824-17

[26] Garcia JL, Patel BK, Ollivier B. Taxonomic, phylogenetic, and ecological diversity of methanogenic archaea. Anaerobe. 2000;**6**:205-226. DOI: 10.1006/anae.2000.034

[27] Liu Y, Whitman WB. Metabolic, phylogenetic, and ecological diversity of the methanogenic archaea. Annals of the New York Academy of Sciences. 2008;**1125**:171-189. DOI: 10.1196/ annals.1419.019

[28] The National Center for Biotechnology Information. Available from: https://www.ncbi.nlm.nih.gov/Taxonomy/Browser/wwwtax.cgi?mode=Tree&id=1235850 [Accessed: March 21, 2025]

[29] Ferry JG. Enzymology of one-carbon metabolism in methanogenic pathways. FEMS Microbiology Reviews. 1999;**23**(1):13-38. DOI: 10.1111/j.1574-6976.1999.tb00390.x

[30] Brugere JF, Borrel G, O'toole PW, Malpuech-brugere C, Alric M. Use of Microorganisms for Reducing the Level of Trimethylamine in a Human Body Cavity, in Particular for the Treatment of Trimethylaminuria or of Bacterial Vaginosis and the Prevention of Cardiovascular Diseases; U.S. Patent No. 10,238,695. Washington, DC: U.S. Patent and Trademark Office; 2019

[31] Ramezani A, Nolin TD, Barrows IR, Serrano MG, Buck GA, Regunathan-Shenk R, et al. Gut colonization with methanogenic archaea lowers plasma trimethylamine N-oxide concentrations in Apolipoprotein e−/− mice. Scientific Reports. 2018;**8**(1):14752. DOI: 10.1038/ s41598-018-33018-5

[32] Brugère JF, Borrel G, Gaci N, Tottey W, O'Toole PW, Malpuech-Brugère C. Archaebiotics: Proposed therapeutic use of archaea to prevent trimethylaminuria and cardiovascular disease. Gut Microbes. 2014;**5**:5-10. DOI: 10.4161/gmic.26749

[33] Whiticar MJ, Faber E, Schoell M. Biogenic methane formation in marine and freshwater environments: CO_2 reduction vs. acetate fermentation-isotope evidence. Geochimica et Cosmochimica Acta. 1986;**50**:693-709

[34] Kubota K, Imachi H, Kawakami S, Nakamura K, Harada H, Ohashi A. Evaluation of enzymatic cell treatments for application of CARD-FISH to methanogens. Journal of Microbiological Methods. 2008;**72**(1):54-59. DOI: 10.1016/j. mimet.2007.10.006

[35] Rauer L, De Tomassi A, Müller CL, Hülpüsch C, Traidl-Hoffmann C, Reiger M, et al. De-biasing microbiome sequencing data: Bacterial morphology-based correction of extraction bias and correlates of chimera formation. Microbiome. 2025;**13**(1):38. DOI: 10.1186/s40168-024-01998-4

[36] Chaudhary PP, Gaci N, Borrel G, O'Toole PW, Brugère JF. Molecular methods for studying methanogens of the human gastrointestinal tract: Current status and future directions. Applied Microbiology and Biotechnology. 2015;**99**(14):5801-5815. DOI: 10.1007/ s00253-015-6739-2

[37] Dridi B, Henry M, El Khéchine A, Raoult D, Drancourt M. High prevalence of *Methanobrevibacter smithii* and *Methanosphaera stadtmanae* detected in the human gut using an improved

DNA detection protocol. PLoS One. 2009;**4**(9):e7063. DOI: 10.1371/journal.pone.0007063

[38] Cisek AA, Bąk I, Stefańska I, Binek M. Selection and optimization of high-yielding DNA isolation protocol for quantitative analyses of methanogenic archaea. Microorganisms. 2022;**10**(3):523. DOI: 10.3390/microorganisms10030523

[39] Pollock J, Glendinning L, Wisedchanwet T, Watson M. The madness of microbiome: Attempting to find consensus "Best practice" for 16S microbiome studies. Applied and Environmental Microbiology. 2018;**84**(7):e02627-e02617. DOI: 10.1128/AEM.02627-17

[40] Tourlousse DM, Narita K, Miura T, Sakamoto M, Ohashi A, Shiina K, et al. Validation and standardization of DNA extraction and library construction methods for metagenomics-based human fecal microbiome measurements. Microbiome. 2021;**9**:95. DOI: 10.1186/s40168-021-01048-3

[41] Bukin YS, Galachyants YP, Morozov IV, Bukin SV, Zakharenko AS, Zemskaya TI. The effect of 16S rRNA region choice on bacterial community metabarcoding results. Scientific Data. 2019;**6**:190007. DOI: 10.1038/sdata.2019.7

[42] Abellan-Schneyder I, Matchado MS, Reitmeier S, Sommer A, Sewald Z, Baumbach J, et al. Primer, pipelines, parameters: Issues in 16S rRNA gene sequencing. mSphere. 2021;**6**(1):e01202-e01220. DOI: 10.1128/mSphere.01202-20

[43] Gilbert JA, Jansson JK, Knight R. The earth microbiome project: Successes and aspirations. BMC Biology. 2014;**12**:69. DOI: 10.1186/s12915-014-0069-1

[44] Human Microbiome Project Consortium. Structure, function and diversity of the healthy human microbiome. Nature. 2012;**486**(7402):207-214. DOI: 10.1038/nature11234

[45] Lin L, Ju F. Evaluation of different 16S rRNA gene hypervariable regions and reference databases for profiling engineered microbiota structure and functional guilds in a swine wastewater treatment plant. Interface Focus. 2023;**13**(4):20230012. DOI: 10.1098/rsfs.2023.0012

[46] Yang B, Wang Y, Qian PY. Sensitivity and correlation of hypervariable regions in 16S rRNA genes in phylogenetic analysis. BMC Bioinformatics. 2016;**17**:135. DOI: 10.1186/s12859-016-0992-y

[47] Sinclair L, Osman OA, Bertilsson S, Eiler A. Microbial community composition and diversity via 16S rRNA gene amplicons: Evaluating the illumina platform. PLoS One. 2015;**10**(2):e0116955. DOI: 10.1371/journal.pone.0116955

[48] Scarano C, Veneruso I, De Simone RR, Di Bonito G, Secondino A, D'Argenio V. The third-generation sequencing challenge: Novel insights for the omic sciences. Biomolecules. 2024;**14**(5):568. DOI: 10.3390/biom14050568

[49] Takai K, Horikoshi K. Rapid detection and quantification of members of the archaeal community by quantitative PCR using fluorogenic probes. Applied and Environmental Microbiology. 2000;**66**(11):5066-5072. DOI: 10.1128/AEM.66.11.5066-5072.2000

[50] Takahashi S, Tomita J, Nishioka K, Hisada T, Nishijima M. Development of a prokaryotic universal primer

for simultaneous analysis of bacteria and archaea using next-generation sequencing. PLoS One. 2014;**9**(8):e105592. DOI: 10.1371/journal.pone.0105592

[51] Caporaso JG, Lauber CL, Walters WA, Berg-Lyons D, Lozupone CA, Turnbaugh PJ, et al. Global patterns of 16S rRNA diversity at a depth of millions of sequences per sample. Proceedings of the National Academy of Sciences of the United States of America. 2011;**108**(Suppl. 1):4516-4522. DOI: 10.1073/pnas.1000080107

[52] Parada AE, Needham DM, Fuhrman JA. Every base matters: Assessing small subunit rRNA primers for marine microbiomes with mock communities, time series and global field samples. Environmental Microbiology. 2016;**18**(5):1403-1414. DOI: 10.1111/1462-2920.13023

[53] Walters W, Hyde ER, Berg-Lyons D, Ackermann G, Humphrey G, Parada A, et al. Improved bacterial 16S rRNA gene (V4 and V4-5) and fungal internal transcribed spacer marker gene primers for microbial community surveys. mSystems. 2015;**1**(1):e00009-e00015. DOI: 10.1128/mSystems.00009-15

[54] Yu MA, Yu TI, Podosokorskaya OA, Toshchakov SV. Analysis of 16S rRNA primer systems for profiling of thermophilic microbial communities. Microbiology. 2019;**88**:671-680. DOI: 10.1134/S0026261719060110

[55] Klindworth A, Pruesse E, Schweer T, Peplies J, Quast C, Horn M, et al. Evaluation of general 16S ribosomal RNA gene PCR primers for classical and next-generation sequencing-based diversity studies. Nucleic Acids Research. 2013;**41**(1):e1. DOI: 10.1093/nar/gks808

[56] Baker GC, Smith JJ, Cowan DA. Review and re-analysis of domain-specific 16S primers. Journal of Microbiological Methods. 2003;**55**(3):541-555. DOI: 10.1016/j.mimet.2003.08.009

[57] Kim JY, Whon TW, Lim MY, Kim YB, Kim N, Kwon MS, et al. The human gut archaeome: Identification of diverse haloarchaea in Korean subjects. Microbiome. 2020;**8**(1):114. DOI: 10.1186/s40168-020-00894-x

[58] Bang C, Schmitz RA. Archaea associated with human surfaces: Not to be underestimated. FEMS Microbiology Reviews. 2015;**39**(5):631-648. DOI: 10.1093/femsre/fuv010

[59] Duller S, Moissl-Eichinger C. Archaea in the human microbiome and potential effects on human infectious disease. Emerging Infectious Diseases. 2024;**30**(8):1505-1513. DOI: 10.3201/eid3008.240181

[60] Thomas CM, Desmond-Le Quéméner E, Gribaldo S, Borrel G. Factors shaping the abundance and diversity of the gut Archaeome across the animal kingdom. Nature Communications. 2022;**13**:3358. DOI: 10.1038/s41467-022-31038-4

[61] Volmer JG, McRae H, Morrison M. The evolving role of methanogenic archaea in mammalian microbiomes. Frontiers in Microbiology. 2023;**14**:1268451. DOI: 10.3389/fmicb.2023.1268451

[62] Freitag TE, Prosser JI. Correlation of methane production and functional gene transcriptional activity in a peat soil. Applied and Environmental Microbiology. 2009;**75**:6679-6687. DOI: 10.1128/AEM.01021-09

[63] Sikolenko MA, Valentovich LN. RiboGrove: A database of full-length prokaryotic 16S rRNA genes derived from completely assembled genomes. Research in Microbiology. 2022;**173**:103936. DOI: 10.1016/j.resmic.2022.103936

[64] Douglas GM, Maffei VJ, Zaneveld JR, Yurgel SN, Brown JR, Taylor CM, et al. PICRUSt2 for prediction of metagenome functions. Nature Biotechnology. 2020;**38**:685-688. DOI: 10.1038/s41587-020-0548-6

[65] Wemheuer F, Taylor JA, Daniel R, Johnston E, Meinicke P, Thomas T, et al. Tax4Fun2: Prediction of habitat-specific functional profiles and functional redundancy based on 16S rRNA gene sequences. Environmental Microbiomes. 2020;**15**:11. DOI: 10.1186/s40793-020-00358-7

[66] Medvedev OS, Ivanova AYU, Belousova MA, Toshchakov SV, Krylova AS, Shirokov IV, et al. Differential effects of carbohydrates on the generation of hydrogen and methane in low- and high-methane-producing rats. In Molecular Hydrogen in Health and Disease; Slezak J, Kura B, Springer Nature Switzerland: Cham. 2024; p. 339-358. ISBN 978-3-031-47375-3

[67] Ben-Amor K, Heilig H, Smidt H, Vaughan EE, Abee T, de Vos WM. Genetic diversity of viable, injured, and dead fecal bacteria assessed by fluorescence-activated cell sorting and 16S rRNA gene analysis. Applied and Environmental Microbiology. 2005;**71**:4679-4689. DOI: 10.1128/AEM.71.8.4679-4689.2005

[68] Peris-Bondia F, Latorre A, Artacho A, Moya A, D'Auria G. The active human gut microbiota differs from the total microbiota. PLoS One. 2011;**6**:e22448. DOI: 10.1371/journal.pone.0022448

[69] Polag D, Keppler F. Global methane emissions from the human body: Past, present and future. Atmospheric Environment. 2019;**214**:1168. DOI: 10.1016/j.atmosenv.2019.116823

[70] Kim JW, Park S, Chung JO, Cho HA, Kim D, Yoon JH, et al. Influencing factors on lactulose breath test results. The Korean Journal of Gastroenterology. 2020;**75**:23-28. DOI: 10.4166/kjg.2020.75.1.23

[71] Santos ANDR, Soares ACF, Oliveira RP, Morais MB. The impact of small intestinal bacterial overgrowth on the growth of children and adolescents. Revista Paulista de Pediatria. 2020;**38**:e2018164. DOI: 10.1590/1984-0462/2020/38/2018164

[72] Le Marchand L, Wilkens LR, Harwood P, Cooney RV. Breath hydrogen and methane in populations at different risk for colon cancer. International Journal of Cancer. 1993;**55**(6):887-890. DOI: 10.1002/ijc.2910550603

[73] Kwiecień J, Hajzler W, Kosek K, Balcerowicz S, Grzanka D, Gościniak W, et al. No correlation between positive fructose hydrogen breath test and clinical symptoms in children with functional gastrointestinal disorders: A retrospective single-centre study. Nutrients. 2021;**13**(8):2891. DOI: 10.3390/nu13082891

[74] Plauzolles A, Uras S, Pénaranda G, Bonnet M, Dukan P, Retornaz F, et al. Small intestinal bacterial overgrowths and intestinal methanogen overgrowths breath testing in a real-life French cohort. Clinical and Translational Gastroenterology. 2023;**14**:e00556. DOI: 10.14309/ctg.0000000000000556

[75] Newberry C, Tierney A, Pickett-Blakely O. Lactulose hydrogen

breath test result is associated with age and gender. BioMed Research International. 2016;**2016**:1064029. DOI: 10.1155/2016/1064029

[76] Kiow LCJ, Bellila R, Therrien A, Sidani S, Bouin M. Predictors of small intestinal bacterial overgrowth in symptomatic patients referred for breath testing. Journal of Clinical Medical Research. 2020;**12**:655-661. DOI: 10.14740/jocmr432

[77] Levitt MD, Furne JK, Kuskowski M, Ruddy J. Stability of human methanogenic flora over 35 years and a review of insights obtained from breath methane measurements. Clinical Gastroenterology and Hepatology. 2006;**4**(2):123-129. DOI: 10.1016/j.cgh.2005.11.006

[78] Peled Y, Gilat T, Liberman E, Bujanover Y. The development of methane production in childhood and adolescence. Journal of Pediatric Gastroenterology and Nutrition. 1985;**4**:575-579

[79] Polag D, Leiß O, Keppler F. Age dependent breath methane in the German population. Science of the Total Environment. 2014;**481**:582-587

[80] Maczulak A, Wolin M, Miller T. Increase in colonic methanogens and total anaerobes in aging rats. Applied and Environmental Microbiology. 1989;**55**:2468-2473

[81] Liu C, Meng Q, Chen Y, Xu M, Shen M, Gao R, et al. Role of age-related shifts in rumen bacteria and methanogens in methane production in cattle. Frontiers in Microbiology. 2017;**8**:1563. DOI: 10.3389/fmicb.2017.01563

[82] Mihajlovski A, Doré J, Levenez F, Alric M, Brugère JF. Molecular evaluation of the human gut methanogenic archaeal microbiota reveals an age-associated increase of the diversity. Environmental Microbiology Reports. 2010;**2**(2):272-280. DOI: 10.1111/j.1758-2229.2009.00116.x

[83] Mohammadzadeh R, Mahnert A, Shinde T, et al. Age-related dynamics of predominant methanogenic archaea in the human gut microbiome. BMC Microbiol. 2025;**25**(1):193. DOI: 10.1186/s12866-025-03921-9

[84] Moissl-Eichinger C, Probst AJ, Birarda G, et al. Human age and skin physiology shape diversity and abundance of archaea on skin. Scientific Reports. 2017;**7**(1):4039. DOI: 10.1038/s41598-017-04197-4

[85] Salas-López M, Vélez-Ixta JM, Rojas-Guerrero DL Piña-Escobedo A, Hernández-Hernández JM, Rangel-Calvillo MN, et al. Human milk archaea associated with neonatal gut colonization and its co-occurrence with bacteria. Microorganisms. 2025;**13**(1):85. DOI: 10.3390/microorganisms13010085

[86] van de Pol JA, van Best N, Mbakwa CA, Thijs C, Savelkoul PH, Arts IC, et al. Gut colonization by methanogenic archaea is associated with organic dairy consumption in children. Frontiers in Microbiology. 2017;**8**:355. DOI: 10.3389/fmicb.2017.00355

[87] Yamabe K, Maeda H, Kokeguchi S, Tanimoto I, Sonoi N, Asakawa S, et al. Distribution of archaea in Japanese patients with periodontitis and humoral immune response to the components. FEMS Microbiology Letters. 2008;**287**(1):69-75. DOI: 10.1111/j.1574-6968.2008.01304.x

[88] Ramiro FS, de Lira E, Soares G, Retamal-Valdes B, Feres M, Figueiredo LC, et al. Effects of different

periodontal treatments in changing the prevalence and levels of archaea present in the subgingival biofilm of subjects with periodontitis: A secondary analysis from a randomized controlled clinical trial. International Journal of Dental Hygiene. 2018;**16**(4):569-575. DOI: 10.1111/idh.12347

[89] Pimentel M, Mayer AG, Park S, Chow EJ, Hasan A, Kong Y. Methane production during lactulose breath test is associated with gastrointestinal disease presentation. Digestive Diseases and Sciences. 2003;**48**:86-92. DOI: 10.1023/A:1021738515885

[90] Kim G, Deepinder F, Morales W, Hwang L, Weitsman S, Chang C, et al. *Methanobrevibacter smithii* is the predominant methanogen in patients with constipation-predominant IBS and methane on breath. Digestive Diseases and Sciences. 2012;**57**(12):3213-3218. DOI: 10.1007/s10620-012-2197-1

[91] Yazici C, Arslan DC, Abraham R, Cushing K, Keshavarzian A, Mutlu EA. Breath methane levels are increased among patients with diverticulosis. Digestive Diseases and Sciences. 2016;**61**:2648-2654. DOI: 10.1007/s10620-016-4174-6

[92] Pimentel M, Chatterjee S, Chow EJ, Park S, Kong Y. Neomycin improves constipation-predominant irritable bowel syndrome in a fashion that is dependent on the presence of methane gas: Subanalysis of a double-blind randomized controlled study. Digestive Diseases and Sciences. 2006;**51**(8):1297-1301. DOI: 10.1007/s10620-006-9104-6

[93] Gandhi A, Shah A, Jones MP, Koloski N, Talley NJ, Morrison M, et al. Methane positive small intestinal bacterial overgrowth in inflammatory bowel disease and irritable bowel syndrome: A systematic review and meta-analysis. Gut Microbes. 2021;**13**(1):1933313. DOI: 10.1080/19490976.2021.1933313

[94] Chumpitazi BP, Weidler EM, Shulman RJ. Lactulose breath test gas production in childhood IBS is associated with intestinal transit and bowel movement frequency. Journal of Pediatric Gastroenterology and Nutrition. 2017;**64**(4):541-545. DOI: 10.1097/MPG.0000000000001295

[95] Lee KN, Lee OY, Koh DH, Sohn W, Lee SP, Jun DW, et al. Association between symptoms of irritable bowel syndrome and methane and hydrogen on lactulose breath test. Journal of Korean Medical Science. 2013;**28**(6):901-907. DOI: 10.3346/jkms.2013.28.6.901

[96] Yokoyama K, Sakamaki A, Takahashi K, Naruse T, Sato C, Kawata Y, et al. Hydrogen-producing small intestinal bacterial overgrowth is associated with hepatic encephalopathy and liver function. PLoS One. 2022;**17**(2):e0264459. DOI: 10.1371/journal.pone.0264459

[97] Djemai K, Gouriet F, Sielezneff I, Mege D, Drancourt M, Grine G. Detection of methanogens in peri-appendicular abscesses: Report of four cases. Anaerobe. 2021;**72**:102470. DOI: 10.1016/j.anaerobe.2021.102470

[98] Takakura W, Oh SJ, Singer-Englar T, Mirocha J, Leite G, Fridman A, et al. Comparing the rates of methane production in patients with and without appendectomy: Results from a large-scale cohort. Scientific Reports. 2020;**10**(1):867. DOI: 10.1038/s41598-020-57662-y

[99] Scanlan PD, Shanahan F, Marchesi JR. Human methanogen diversity and incidence in healthy and

diseased colonic groups using mcrA gene analysis. BMC Microbiology. 2008;**8**:79. DOI: 10.1186/1471-2180-8-79

[100] Million M, Tidjani Alou M, Khelaifia S, Bachar D, Lagier JC, Dione N, et al. Increased gut redox and depletion of anaerobic and methanogenic prokaryotes in severe acute malnutrition. Scientific Reports. 2016;**6**:26051. DOI: 10.1038/srep26051. Retraction in: Sci Rep. 2023 Oct 30;13(1):18589. DOI: 10.1038/s41598-023-44597-3

[101] Armougom F, Henry M, Vialettes B, Raccah D, Raoult D. Monitoring bacterial community of human gut microbiota reveals an increase in lactobacillus in obese patients and methanogens in anorexic patients. PLoS One. 2009;**4**(9):e7125. DOI: 10.1371/journal.pone.0007125

[102] Rahman MM, Islam F, Or-Rashid MH, Mamun AA, Rahaman MS, Islam MM, et al. The gut microbiota (microbiome) in cardiovascular disease and its therapeutic regulation. Frontiers in Cellular and Infection Microbiology. 2022;**12**:1-22. DOI: 10.3389/fcimb.2022.903570

[103] Chen L, Zhernakova DV, Kurilshikov A, Andreu-Sánchez S, Wang D, Augustijn HE, et al. Influence of the microbiome, diet and genetics on inter-individual variation in the human plasma metabolome. Nature Medicine. 2022;**28**(11):2333-2343. DOI: 10.1038/s41591-022-02014-8

[104] Grammatopoulos K, Antoniou VD, Mavrothalassitis E, Mouziouras D, Argyris AA, Emmanouil E, et al. Association of gut microbiota composition and their metabolites with subclinical atheromatosis: A systematic review. American Heart Journal Plus. 2022;**23**:100219. DOI: 10.1016/j.ahjo.2022.100219

[105] Chakaroun RM, Olsson LM, Bäckhed F. The potential of tailoring the gut microbiome to prevent and treat cardiometabolic disease. Nature Reviews. Cardiology. 2023;**20**(4):217-235. DOI: 10.1038/s41569-022-00771-0. Epub 2022 Oct 14

[106] Zhu W, Gregory JC, Org E, Buffa JA, Gupta N, Wang Z, et al. Gut microbial metabolite TMAO enhances platelet hyperreactivity and thrombosis risk. Cell. 2016;**165**(1):111-124. DOI: 10.1016/j.cell.2016.02.011. Epub 2016 Mar 10

[107] Dolkar P, Deyang T, Anand N, Rathipriya AG, Hediyal TA, Chandrasekaran V, et al. Trimethylamine-N-oxide and cerebral stroke risk: A review. Neurobiology of Disease. 2024;**192**:106423. DOI: 10.1016/j.nbd.2024.106423. Epub 2024 Jan 28

[108] Garcia-Fernandez H, Alcala-Diaz JF, Quintana-Navarro GM, Lopez-Moreno J, Luque-Cordoba D, Ruiz-Diaz Narvaez E, et al. Trimethylamine oxidation into the proatherogenic trimethylamine N-oxide is higher in coronary heart disease men: From the CORDIOPREV Study. World Journal of Men's Health. 2025;**43**(1):249-258. DOI: 10.5534/wjmh.230366

[109] Awosika AO, Anastasopoulou C. Trimethylaminuria. Treasure Island (FL): StatPearls Publishing; 2025

[110] Fadhlaoui K, Arnal ME, Martineau M, et al. Archaea, specific genetic traits, and development of improved bacterial live biotherapeutic products: Another face of next-generation probiotics. Applied Microbiology and Biotechnology. 2020;**104**(11):4705-4716. DOI: 10.1007/s00253-020-10599-8

[111] Neill AR et al. Conversion of choline methyl groups through trimethylamine into methane in the rumen. The Biochemical Journal. 1978;**170**:529-535

[112] Borrel G, McCann A, Deane J, et al. Genomics and metagenomics of trimethylamine-utilizing archaea in the human gut microbiome. The ISME Journal. 2017;**11**(9):2059-2074. DOI: 10.1038/ismej.2017.72

[113] de la Cuesta-Zuluaga J, Spector TD, Youngblut ND, Ley RE. Genomic insights into adaptations of trimethylamine-utilizing methanogens to diverse habitats, including the human gut. mSystems. 2021;**6**(1):e00939-e00920. DOI: 10.1128/mSystems.00939-20

[114] Liu Y, Chu M, Wang D, Luo Y, Liu Z, Zhao J. Risk factors for small intestinal bacterial overgrowth in patients with acute ischaemic stroke. Journal of Medical Microbiology. 2023;**72**(2):10.1099/jmm.0.001666. DOI: 10.1099/jmm.0.001666

[115] Takakura W, Chang C, Pimentel M, Mo G, Torosyan J, Hosseini A, et al. Exhaled methane is associated with a lower heart rate. Cardiology. 2022;**147**(2):225-229. DOI: 10.1159/000521434. Epub 2021 Dec 13

[116] Mollar A, Villanueva MP, Núñez E, Carratalá A, Mora F, Bayés-Genís A, et al. Hydrogen- and methane-based breath testing and outcomes in patients with heart failure. Journal of Cardiac Failure. 2019;**25**(5):319-327. DOI: 10.1016/j.cardfail.2018.10.004. Epub 2018 Oct 19

[117] Song Y, Liu Y, Qi B, Cui X, Dong X, Wang Y, et al. Association of small intestinal bacterial overgrowth with heart failure and its prediction for short-term outcomes. Journal of the American Heart Association. 2021;**10**(7):e015292. DOI: 10.1161/JAHA.119.015292

[118] Smirnova MD, Svirida ON, Blankova ZN, Vitsenya MV, Ivanova AY, Ageev FT, et al. Influence of enzymatic activity of microbiota on hemodynamics, arterial stiffness and emotional state of patients with mild-to-moderate risk of cardiovascular complications. Russian Cardiology Bulletin. 2024;**19**(4-2):129-137. DOI: 10.17116/Cardiobulletin202419042129. (In Russ.)

[119] Bhute SS, Suryavanshi MV, Joshi SM, Yajnik CS, Shouche YS, Ghaskadbi SS. Gut microbial diversity assessment of Indian type-2-diabetics reveals alterations in eubacteria, archaea, and eukaryotes. Frontiers in Microbiology. 2017;**8**:214. DOI: 10.3389/fmicb.2017.00214

[120] Mathur R, Goyal D, Kim G, Barlow GM, Chua KS, Pimentel M. Methane-producing human subjects have higher serum glucose levels during oral glucose challenge than non-methane producers: A pilot study of the effects of enteric methanogens on glycemic regulation. Research Journal of Endocrinology and Metabolism. 2014;**2**:2. DOI: 10.7243/2053-3640-2-2

[121] Ouyang Q, Xu Y, Ban Y, Li J, Cai Y, Wu B, et al. Probiotics and prebiotics in subclinical hypothyroidism of pregnancy with small intestinal bacterial overgrowth. Probiotics and Antimicrobial Proteins. 2024;**16**(2):579-588. DOI: 10.1007/s12602-023-10068-4

[122] Zhang H, DiBaise JK, Zuccolo A, Kudrna D, Braidotti M, Yu Y, et al. Human gut microbiota in obesity and after gastric bypass. Proceedings of the National Academy of Sciences of the United States of America. 2009;**106**(7):2365-2370. DOI: 10.1073/pnas.0812600106

[123] Basseri RJ, Basseri B, Pimentel M, Chong K, Youdim A, Low K, et al. Intestinal methane production in obese individuals is associated with a higher body mass index. Gastroenterology Hepatology (N Y). 2012;**8**(1):22-28

[124] Million M, Maraninchi M, Henry M, Armougom F, Richet H, Carrieri P, et al. Obesity-associated gut microbiota is enriched in *Lactobacillus reuteri* and depleted in *Bifidobacterium animalis* and *Methanobrevibacter smithii*. International Journal of Obesity. 2012;**36**(6):817-825. DOI: 10.1038/ijo.2011.153. Epub 2011 Aug 9. Retraction in: Int J Obes (Lond). 2024;48(10):1516. DOI: 10.1038/s41366-024-01595-3

[125] DiBaise JK, Zhang H, Crowell MD, Krajmalnik-Brown R, Decker GA, Rittmann BE. Gut microbiota and its possible relationship with obesity. Mayo Clinic Proceedings. 2008;**83**:460-469. DOI: 10.4065/83.4.460

[126] Mathur R, Amichai M, Chua KS, Mirocha J, Barlow GM, Pimentel M. Methane and hydrogen positivity on breath test is associated with greater body mass index and body fat. The Journal of Clinical Endocrinology and Metabolism. 2013;**98**:E698-E702. DOI: 10.1210/jc.2012-3144

[127] Cortez APB, Fisberg M, de Morais MB. Intestinal permeability and small intestine bacterial overgrowth in excess weight adolescents. Pediatric Obesity. 2021;**16**:e12741. DOI: 10.1111/ijpo.12741

[128] Arenas-Montes J, Perez-Martinez P, Vals-Delgado C, Romero-Cabrera JL, Cardelo MP, Leon-Acuña A, et al. Owning a pet is associated with changes in the composition of gut microbiota and could influence the risk of metabolic disorders in humans. Animals (Basel). 2021;**11**(8):2347. DOI: 10.3390/ani11082347

[129] Misnikova I, Kovaleva Y, Shokur S, LeBaron TW, Povarova O, Medvedev O. Hydrogen and methane detection in breath in response to two different types of dietary fiber and its relationship to postprandial glucose concentration in obese patients with type 2 diabetes and normoglycemic subjects. Nutrients. 2025;**17**(5):917. DOI: 10.3390/nu17050917

[130] Tremlett H, Bauer KC, Appel-Cresswell S, Finlay BB, Waubant E. The gut microbiome in human neurological disease: A review. Annals of Neurology. 2017;**81**(3):369-382. DOI: 10.1002/ana.24901. Epub 2017 Mar 20

[131] Vuotto C, Battistini L, Caltagirone C, Borsellino G. Gut microbiota and disorders of the central nervous system. The Neuroscientist. 2020;**26**(5-6):487-502. DOI: 10.1177/1073858420918826. Epub 2020 May 22

[132] Romano S, Savva GM, Bedarf JR, Charles IG, Hildebrand F, Narbad A. Meta-analysis of the Parkinson's disease gut microbiome suggests alterations linked to intestinal inflammation. NPJ Parkinson's Disease. 2021;**7**(1):27. DOI: 10.1038/s41531-021-00156-z

[133] Fumagalli A, Castells-Nobau A, Trivedi D, Garre-Olmo J, Puig J, Ramos R, et al. Archaea methanogens are associated with cognitive performance through the shaping of gut microbiota, butyrate and histidine metabolism. Gut Microbes. 2025;**17**(1):2455506. DOI: 10.1080/19490976.2025.2455506

[134] Roccarina D, Lauritano EC, Gabrielli M, Franceschi F, Ojetti V, Gasbarrini A. The role of methane in intestinal diseases. The American Journal of Gastroenterology. 2010;**105**(6):1250-1256. DOI: 10.1038/ajg.2009.744. Epub 2010 Mar 9

[135] Coker OO, Wu WKK, Wong SH, Sung JJY, Yu J. Altered gut archaea composition and interaction with bacteria are associated with colorectal cancer. Gastroenterology. 2020;**159**(4):1459-1470.e5. DOI: 10.1053/j.gastro.2020.06.042. Epub 2020 Jun 20

[136] Alonso-Sáez L, Waller AS, Mende DR, Bakker K, Farnelid H, Yager PL, et al. Role for urea in nitrification by polar marine archaea. Proceedings of the National Academy of Sciences of the United States of America. 2012;**109**(44):17989-17994. DOI: 10.1073/pnas.1201914109. Epub 2012 Oct 1

[137] Cai M, Kandalai S, Tang X, Zheng Q. Contributions of human-associated archaeal metabolites to tumor microenvironment and carcinogenesis. Microbiology Spectrum. 2022;**10**(2):e0236721. DOI: 10.1128/spectrum.02367-21. Epub 2022 Feb 28

[138] Vierbuchen T, Bang C, Rosigkeit H, Schmitz RA, Heine H. The human-associated archaeon *Methanosphaera stadtmanae* is recognized through its RNA and induces TLR8-dependent NLRP3 inflammasome activation. Frontiers in Immunology. 2017;**8**:1535

[139] Oka S, Watanabe M, Ito E, et al. Archaeal glycerolipids are recognized by C-type lectin receptor mincle. Journal of the American Chemical Society. 2023;**145**(33):18538-18548. DOI: 10.1021/jacs.3c05473

[140] Kuehnast T, Kumpitsch C, Mohammadzadeh R, Weichhart T, Moissl-Eichinger C, Heine H. Exploring the human archaeome: Its relevance for health and disease, and its complex interplay with the human immune system. The FEBS Journal. 2024;**292**(6):1316-1329. DOI: 10.1111/febs.17123

[141] Yamabe K, Maeda H, Kokeguchi S, Soga Y, Meguro M, Naruishi K, et al. Antigenic group II chaperonin in *Methanobrevibacter oralis* may cross-react with human chaperonin CCT. Molecular Oral Microbiology. 2010;**25**:112-122. DOI: 10.1111/j.2041-1014.2009.00548.x

[142] Blais Lecours P, Marsolais D, Cormier Y, Berberi M, Haché C, Bourdages R, et al. Increased prevalence of *Methanosphaera stadtmanae* in inflammatory bowel diseases. PLoS One. 2014;**9**(2):e87734. DOI: 10.1371/journal.pone.0087734

[143] Blais Lecours P, Duchaine C, Taillefer M, Tremblay C, Veillette M, Cormier Y, et al. Immunogenic properties of archaeal species found in bioaerosols. PLoS One. 2011;**6**(8):e23326. DOI: 10.1371/journal.pone.0023326

[144] Bernatchez E, Gold MJ, Langlois A, Blais-Lecours P, Boucher M, Duchaine C, et al. *Methanosphaera stadtmanae* induces a type IV hypersensitivity response in a mouse model of airway inflammation. Physiological Reports. 2017;**5**(7):e13163. DOI: 10.14814/phy2.13163

[145] Bang C, Weidenbach K, Gutsmann T, Heine H, Schmitz RA. The intestinal archaea *Methanosphaera stadtmanae* and *Methanobrevibacter smithii* activate human dendritic cells. PLoS One. 2014;**9**(6):e99411. DOI: 10.1371/journal.pone.0099411

[146] Bang C, Vierbuchen T, Gutsmann T, Heine H, Schmitz RA. Immunogenic properties of the human gut-associated archaeon *Methanomassiliicoccus luminyensis* and its susceptibility to antimicrobial peptides. PLoS One. 2017;**12**(10):e0185919. DOI: 10.1371/journal.pone.0185919

[147] Gottlieb K, Wacher V, Sliman J, Pimentel M. Review article: Inhibition of methanogenic archaea by statins as a targeted management strategy for constipation and related disorders. Alimentary Pharmacology & Therapeutics. 2016;**43**(2):197-212. DOI: 10.1111/apt.13469

[148] Muskal SM, Sliman J, Kokai-Kun J, Pimentel M, Wacher V,

Gottlieb K. Lovastatin lactone may improve irritable bowel syndrome with constipation (IBS-C) by inhibiting enzymes in the archaeal methanogenesis pathway. F1000Res. 2016;**5**:606. DOI: 10.12688/f1000research.8406.3

[149] Halmos EP, Power VA, Shepherd SJ, Gibson PR, Muir JG. A diet low in FODMAPs reduces symptoms of irritable bowel syndrome. Gastroenterology. 2014;**146**(1):67-75.e5. DOI: 10.1053/j.gastro.2013.09.046

[150] Ojetti V, Petruzziello C, Migneco A, Gnarra M, Gasbarrini A, Franceschi F. Effect of *Lactobacillus reuteri* (DSM 17938) on methane production in patients affected by functional constipation: A retrospective study. European Review for Medical and Pharmacological Sciences. 2017;**21**(7):1702-1708. DOI: 10.3390/medicina57070643

[151] Gerber PL, Steinfeld H, Henderson B, Mottet A, Opio C, Dijkman J, et al. Tackling Climate Change through Livestock—A Global Assessment of Emissions and Migration Opportunities. Rome, Italy: Food and Agricultural Organizations of the United Nations (FAO); 2013. Available from: https://www.fao.org/3/i3437e/i3437e00.htm [Accessed: March 24, 2025]

[152] Johnson KA, Johnson DE. Methane emissions from cattle. Journal of Animal Science. 1995;**73**:2483-2492. DOI: 10.2527/1995.7382483x

[153] Misiukiewicz A, Gao M, Filipiak W, Cieslak A, Patra AK, Szumacher-Strabel M. Review: Methanogens and methane production in the digestive systems of nonruminant farm animals. Animal: An International Journal of Animal Bioscience. 2021;**15**(1):100060. DOI: 10.1016/j.animal.2020.100060

[154] Poulsen M, Schwab C, Jensen BB, Engberg RM, Spang A, Canibe N, et al. Methylotrophic methanogenic thermoplasmata implicated in reduced methane emissions from bovine rumen. Nature Communications. 2013;**4**:1428. DOI: 10.1038/ncomms2432

[155] Patra AK, Park T, Kim M, Yu Z. Rumen methanogens and mitigation of methane emission by anti-methanogenic compounds and substances. Journal of Animal Science and Biotechnology. 2017;**2017**(8):13. DOI: 10.1186/s40104-017-0145-9

[156] Li QS, Wang R, Ma ZY, Zhang XM, Jiao JZ, Zhang ZG, et al. Dietary selection of metabolically distinct microorganisms drives hydrogen metabolism in ruminants. The ISME Journal. 2022;**16**(11):2535-2546. DOI: 10.1038/s41396-022-01294-9

[157] Denman SE, Tomkins NW, McSweeney CS. Quantitation and diversity analysis of ruminal methanogenic populations in response to the antimethanogenic compound bromochloromethane. FEMS Microbiology Ecology. 2007;**62**:313-322. DOI: 10.1111/j.1574-6941.2007.00394.x

[158] Glasson CRK, Kinley RD, de Nys R, King N, Adams SL, Packer MA, et al. Benefits and risks of including the bromoform containing seaweed *Asparagopsis* in feed for the reduction of methane production from ruminants. Algal Research. 2022;**64**:102673. DOI: 10.1016/j.algal.2022.102673

[159] Duin EC, Wagner T, Shima S, Prakash D, Cronin B, Yáñez-Ruiz DR, et al. Mode of action uncovered for the specific reduction of methane emissions from ruminants by the small molecule 3-nitrooxypropanol. Proceedings of the National Academy of Sciences of the United States of America.

2016;**113**:6172-6177. DOI: 10.1073/pnas.1600298113

[160] Vyas D, McGinn SM, Duval SM, Kindermann M, Beauchemin KA. Effects of sustained reduction of enteric methane emissions with dietary supplementation of 3-nitrooxypropanol on growth performance of growing and finishing beef cattle. Journal of Animal Science. 2016;**94**:2024-2034. DOI: 10.2527/jas.2015-0268

[161] Almeida AK, Cowley F, McMeniman JP, Karagiannis A, Walker N, Tamassia LFM, et al. Effect of 3-nitrooxypropanol on enteric methane emissions of feedlot cattle fed with a tempered barley-based diet with canola oil. Journal of Animal Science. 2023;**101**:skad237. DOI: 10.1093/jas/skad237

[162] Marounek M, Suchorska O, Savka O. Effect of substrate and feed antibiotics on in vitro production of volatile fatty acids and methane in caecal contents of chickens. Animal Feed Science and Technology. 1999;**80**:223-230. DOI: 10.1016/S0377-8401(99)00065-6

[163] Karekar S, Stefanini R, Ahring B. Homo-acetogens: Their metabolism and competitive relationship with hydrogenotrophic methanogens. Microorganisms. 2022;**10**(2):397. DOI: 10.3390/microorganisms10020397

[164] Whitelaw FG, Eadie JM, Bruce LA, Shand WJ. Methane formation in faunated and ciliate-free cattle and its relationship with rumen volatile fatty acid proportions. The British Journal of Nutrition. 1984;**52**:261-275. DOI: 10.1079/BJN19840094

[165] Gong YL, Liang JB, Jahromi MF, Wu YB, Wright AG, Liao XD. Mode of action of *Saccharomyces cerevisiae* in enteric methane mitigation in pigs. Animal. 2018;**12**:239-245. DOI: 10.1017/S1751731117001732

[166] Pinares-Patiño CS, Hickey SM, Young EA, Dodds KG, MacLean S, Molano G, et al. Heritability estimates of methane emissions from sheep. Animal. 2013;**7**:316-321. DOI: 10.1017/S1751731113000864

[167] Baca-González V, Asensio-Calavia P, González-Acosta S, Pérez de la Lastra JM, Morales de la Nuez A. Are vaccines the solution for methane emissions from ruminants? A systematic review. Vaccines. 2020;**8**(3):460. DOI: 10.3390/vaccines8030460

Section 4

Therapeutic Potential of New Metabiotics and Probiotics

Chapter 7

Multicomponent Metabiotic Actoflor-S Has Therapeutic Potential for Bowel, Liver, and Metabolic Disorders

Timur Vakhitov and Stanislav Sitkin

Abstract

One of the aspects of bacterial symbiosis with the host is the exchange of metabolites. Stimulation of the growth of beneficial bacteria in the microbiota using the metabiotic Actoflor®-S led to a change in the composition of the flow of its metabolites. Analysis of the blood metabolome of mice by GC-MS showed that the animals receiving Actoflor®-S had an increased concentration of almost all amino acids and a decreased concentration of most carboxylic acids and their derivatives, including lactic acid. Comparison of the obtained results with literature data showed that the action of Actoflor®-S leads to normalization of the concentration of at least 10 marker compounds, whose concentration was significantly changed in patients with ulcerative colitis. The property to normalize the composition of the metabolome turned out to be inherent only to the entire set of Actoflor®-S components, but not to its parts, which is an indicator of the emergent properties of the metabiotic. Overall, the results of the work also indicate that all metabolites of the metabolome are important participants in maintaining health; often, they have individual activity, but in most cases, their activity is a consequence of their combined action.

Keywords: Actoflor®-S, metabiotics, microbiota, growth stimulants, composition modulation, interaction with the host, laboratory mice, blood metabolome, GC-MS, disease markers, biological activity, antagonistic activity

1. Introduction

The development of the metabiotic Actoflor®-S is based on fundamental research in the field of growth physiology of microorganism cultures, or to be more precise, in the field of autoregulation of growth of these cultures. An important issue that microbial physiologists tried to resolve in the late 19th and early 20th centuries was the question of why microorganism cultures began to grow after a certain lag period and why their growth stops even if there are sufficient nutrients and oxygen in the environment.

Even before the general interest in these issues arose, Louis Pasteur wrote in his work "Memoirs on Alcoholic Fermentation" [1], that yeast is unable to grow in pure or sugar water when a small number of cells are introduced and restores this ability if the seeding density is increased. Pasteur saw the reason for this in the lack of certain nitrogenous and mineral substances secreted into the medium by the cells themselves. The effect discovered by Pasteur was later called the effect of autostimulation of growth.

The systematic study of the effect of autostimulation of growth discovered by Pasteur began in 1901 with the work of a third-year student at the University of Leuven (Belgium), Wildiers [2]. He worked under the supervision of Professor M. Ide, but since the article was presented at a student paper competition, it was published only under the name of Wildiers. Like Pasteur, Ide and Wildiers discovered that after seeding, cells secrete certain substances into the medium that promote their growth and increase the fermentation rate but are not the main sources of nitrogen and carbon. These substances were given the provisional name "bios", which was supposed to be replaced by a chemical name after the composition of "bios" was established.

Similar results to those of Wildiers were obtained by the Russian scientist Ya. Nikitinsky, who, like Wildiers and Pasteur, showed (1904) that yeast and mold fungi, during the growth process, release substances capable of activating the reproduction of several microorganisms [3].

Subsequent work led to an important conclusion about the complex structure of "bios." In 1928, Eastcott determined that the active part of one of the fractions ("bios I") was the long-known substance mesoinositol (hexahydroxycyclohexane). In 1936, Fritz Kögl, together with Tönnies, isolated the most active component of bios ("bios II") and called it biotin, and Miller showed that some amino acids should also be included in "bios." Later, β-alanine and pantothenic acid were included in the "bios," and then vitamin B6, para-aminobenzoic, and folic acids [3].

Yeast is a eukaryotic microorganism. Several years after Wildiers' work, Otto Rahn [4] conducted similar experiments on prokaryotic microorganisms, bacteria. He also concluded that the growth rate of bacteria is determined by the concentration of a certain heat-stable substance that they form and secrete into the growth medium. The higher the concentration of bacteria after seeding, the faster this substance accumulated in the medium, and, as a result, the growth of bacteria began.

The identification of these compounds, however, was delayed for almost a century [5], and their search was not marked by such significant events as the discovery of vitamins. One of the attempts to identify autostimulators of bacterial growth was undertaken by the group of Lankford and co-authors in the second half of the twentieth century. The authors did not know about the experiments of O. Rahn but confirmed the presence and functions of autostimulators of growth and also came to the conclusion that these compounds are necessary not only for the initiation of growth but also for its maintenance throughout the process.

Almost 10 years later (1974–1975), the effect of autostimulation of growth was studied by Russian scientists Korovina and co-authors (in experiments with Clostridia), as well as Pshenichnov and Barikhin and co-authors (*E. coli* M-17) [5]. The latter authors, in particular, discovered that the concentration of autostimulators in the bacterial growth medium did not change monotonically but periodically; after a regular increase during the lag period, it decreased and then increased again. There could be several such fluctuations during the growth process, and an increase in activity was accompanied by bacterial proliferation, while when it fell, proliferation slowed down. The presence of fluctuations was explained by the fact that under the growth conditions used by the authors, the periodic culture retained some synchronicity of cell division [5].

Fluctuations in stimulating (inhibitory) activity were also observed after the end of growth and were accompanied by corresponding fluctuations in the number of bacteria, which was first shown in 1960 by Harrison in experiments with *K. aerogenes* and then (1973) confirmed by Sazonova, Weissman, and Korovina in experiments with Clostridia and a group of coworkers led by Pshenichnov (1975) in experiments with *E. coli* M-17 [5]. Fluctuations could arise spontaneously or after adding culture medium (CM) filtrates to the medium.

Many Russian researchers, for example, Shilin et al. in 1983, Pechurkin and Brilkov in 1984, and Pechurkin et al. in 1990, in their work on continuous bacterial cultures, noted that autostimulators are also necessary for growth. Since a denser, continuous culture secreted more autostimulators, its growth rate was higher than that of a similar, less dense culture [5].

All the above-mentioned works contributed significantly to the study of the effect of autostimulation on growth, but their results often contradicted each other. Autostimulators appeared as some elusive compounds, the identification of which was always expected but for some reason was constantly postponed.

The last unsuccessful attempt to identify autostimulators of bacterial growth was undertaken in the laboratory of the famous British biochemist Douglas Kell [6]. After successfully completing a number of stages of isolating autostimulators, the authors nevertheless lost the desired activity, presumably because they did not have reliable methods for testing it and did not take into account the possibility of the combined synergistic action of several metabolites [5].

Our work in this area began in the first half of the 1980s and was aimed at studying the processes of autoregulation in the bacterium *E. coli* M-17, isolated from patients by the Russian scientist L. G. Peretz, who developed a number of probiotic preparations based on it, starting with coli-yogurt and ending with a lyophilized form of the drug "colibacterin."

When planning this study, it seemed tempting to us to isolate and identify substances that increase the growth rate, antagonistic activity, and other beneficial properties of *E. coli* M-17 to increase its therapeutic and prophylactic potential.

First, we were convinced that the *E. coli* M-17 culture medium contains growth autostimulators, learned to determine their activity in several mutually complementary ways, and obtained culture media with activity much higher than their activity during conventional cultivation. When fractioning the autostimulators, it was confirmed that the highly active culture medium contained several active fractions, the combined effect of which significantly exceeded the effect of each of them. In the low-activity medium, the autostimulators themselves and their varieties were significantly fewer.

In the course of further work, almost all autostimulators of this strain were identified, and a composition was created on their basis, which was called Actoflor®-S (activator of microflora-Synthetic). Further studies have shown that it is capable of stimulating the growth and antagonistic activity not only of *E. coli* M-17 but also of other well-known probiotic microorganisms, especially bifidobacteria and lactobacilli.

Although a number of components of Actoflor®-S (e.g., succinic and glutamic acids, methionine, and lysine) had pronounced individual biological activity, their combined action was stronger and more specific. Other components acted only together (e.g., leucine, valine, and alanine). Lactic, acetic, and formic acids (in the form of salts) were, on the one hand, antagonism stimulators and, on the other hand, density-dependent autoregulators stimulating the growth of probiotic *E. coli*, *Lactobacilli*, *Bifidobacteria*, and other beneficial bacteria when their numbers in the human microbiota were reduced and limiting their growth in cases of excessive reproduction of these bacteria.

Currently, the metabiotic Actoflor®-S is produced by the pharmaceutical company Solopharm in three modifications. The first of them is Actoflor®-S, a liquid form for adults, somewhat similar to the well-known drug Hilak-forte, but, unlike it, it does not contain any unidentified substances, including any high-molecular compounds—proteins, nucleic acids, lipopolysaccharides, and other substances that are potential allergens in the composition of all drugs of microbial origin. Actoflor®-duo capsules differ in that they additionally contain an important autoregulation factor, the amino acid cysteine, which is unstable in the liquid form of Actoflor®-S, as well as riboflavin and folic acid (vitamins B2 and B9), which are among the biologically active substances supplied to the host by commensal *E. coli* M-17. Lactic acid is included in Actoflor®-duo in the form of calcium lactate; i.e., the entire complex is additionally a source of easily digestible calcium. The capsule form has a longer shelf life and ensures the delivery of biologically active substances directly to the site of action—the large intestine.

The Actoflor®-kids form of the drug deserves special attention; it has been adapted by the employees of the pharmaceutical company Solopharm specifically for children starting from the age of three. The developers of the children's form relied on the fact that it is at this age that the formation of intestinal microbiota is completed, and, therefore, its timely and correct formation allows you to avoid many troubles of adolescence and adulthood associated with the need to correct the composition of the microbiota after the completion of the main stages of its formation.

All components in all three forms of the drug are pure chemicals produced by well-known pharmaceutical companies, guaranteeing a high degree of purification, and can be used in dietary supplements and/or pharmaceuticals.

2. The effect of the metabiotic Actoflor®-S and its components on the composition of the host blood metabolome

The mechanism of action of all three forms of the metabiotic Actoflor® can be formulated as a number of provisions:

1. The metabiotic serves as a source of importance, including regulatory, microbial metabolites that the host receives from the microbiota in insufficient quantities due to a violation of its composition.

2. Regulatory metabolites that are part of the metabiotic affect the composition and biological activity of the microbiota, which performs immune, trophic, psychoneurological, and a number of other functions in relation to the host organism.

3. Under the influence of the metabiotic, there is a change in the qualitative and quantitative composition of the metabolites that the host receives from the microbiota, resulting from a change in the composition of the microbiota itself.

4. Metabolites of the metabiotic and microbiota cause a response from the host in the form of a change in the composition and quantity of mucin and metabolites that the host supplies to the microbiota. The host's response, combined with the action of the metabiotic, entails further changes in the composition and functional activity of the microbiota, as a result of which the entire microbiota-host system moves to a new level of self-consistent development.

Earlier, in our experiments and experiments of other authors with bacterial cultures, eukaryotic cell cultures, organotypic tissue cultures, as well as works aimed at identifying the features of the metabolome in various diseases, it was shown that blood metabolome analysis is an informative tool for studying the specific features of the disease, the action of drugs, dietary supplements, and other effects on the host [5].

The purpose of this experimental study was to continue these works, aimed at identifying the features of the effect of the dietary supplement Actoflor®-S and its components on the composition of the host metabolome. The advantage of the metabolomic analysis method is that it allows us to judge the nature of the action of drugs at the level of the entire host organism and its microbiota.

3. Materials and methods

Actoflor®-S, a biologically active additive solution for oral administration, is manufactured by Grotex LLC, Solopharm, https://actoflor.ru/en. Composition of BAA Actoflor®-S in g /l: L-aspartic acid (0.91), glycine (1.20), L-leucine (1.50), L-alanine (2.03), L-valine (6.68), L-glutamic acid (6.71), L-lysine hydrochloride (8.33), L-methionine (3.4), sodium formate (15.5), succinic acid (16.2), D, L-lactic acid (24.7), sodium acetate (116.0). The carboxylic acid complex (CAC) included the last four components of Actoflor®-S in the same concentrations. The concentration of the sodium acetate solution was 116 g/l. All components were provided by Grotex LLC, Solopharm. For administration to animals, the preparation and stock solutions of CAC and acetate were diluted 100 times before use and administered to animals intragastrically in a volume of 200 μl/mouse.

The study involved outbred white mice, females, weighing approximately 18 g. (Rappolovo Nursery, St. Petersburg, Russian Federation). All animals were divided into four groups of six animals. The control group received saline; the experimental groups received Actoflor®-S, CAC, and sodium acetate. The drugs were administered for 3 days under standard feeding conditions; in the evening of the 3rd day, the animals were stopped feeding, and in the morning of the 4th day, the drugs were administered on an empty stomach. After 3 hours, the animals were sacrificed by decapitation, and their blood was collected. Blood serum obtained by the standard method (with the addition of heparin) was frozen at −18°C.

For metabolomic analysis, blood serum samples were defrosted at room temperature. 0.5 ml of acetonitrile was added to 0.15 ml; the mixture was stirred for 3 minutes; the clear supernatant was collected; and the sediment was washed with 0.5 ml of acetonitrile. Both solutions were combined, an internal standard (10 μl of trideuteromethanol ester of tridecanoic acid dissolved in methanol (2 mg/ml)) was added, and the mixture was dried in a nitrogen stream. 100 μl of the background of the silylating agent (BSTFA) (N,O-bis(trimethylsilyl)-trifluoroacetamid) was added to the residue, and the mixture was kept at 50°C for 3 minutes. The obtained samples were analyzed on a GCMS-QP2010 Plus chromatograph mass spectrometer from SHIMADZU with GCMS solution software for data processing and the NIST08 mass spectral information database.

Chromatographic separation was performed on an Ultra-2 capillary column 25 m long and 0.2 mm in internal diameter. The chromatogram of each sample was recorded in two modes: (1) From 1 minute to 6.4 minutes, monitoring of ions m/z = 103, 117, 145 (SIM mode); (2) from 6.4 minutes to 40 minutes by the total ion current in the mass range from 35 to 550 (SCAN mode). The choice of the first recording mode was due to the need to record the peaks of formic, acetic, and butyric acids against the background of the silylating agent (BSTFA), contained in excess.

The compounds were identified by comparing their mass spectra with the mass spectra contained in the NIST08 computer library, as well as based on the retention times of the compounds and our spectral information. To obtain quantitative data, the peak areas of the identified components were expressed as a percentage of the sum of the areas of all identified compounds.

Statistical processing of the results was performed using standard statistical programs. To assess statistical significance, the Mann-Whitney, Kolmogorov-Smirnov, and Wilcoxon paired tests were used.

4. Results

According to the chromatograph mass spectrometric analysis, about 100 peaks of compounds with retention times from 1.77 to 31.6 minutes were identified in the blood serum. Almost all of the identified compounds were trimethylsilyl derivatives of blood metabolites of laboratory animals. Some blood metabolites formed not one, but several silyl derivatives by different functional groups or by two or more functional groups simultaneously. As a result, the number of identified metabolites was less than the number of silyl derivatives and amounted to 88 different compounds. Studies with the drug Actoflor®-S, in contrast to its component "acetate" and the mixture "CAC," were carried out twice with an interval of 1 year and showed good reproducibility and comparability of the results. This explains their more complete representation and the absence of some compounds in the groups "Acetate" and "CAC." Comparison of the results of the two studies made it possible to clarify a number of conclusions and increase their reliability.

All identified compounds were divided into six groups (**Tables 1–6**). The first five groups contained 51 compounds (58% of all compounds), the concentration of which in the blood of laboratory animals had at least one statistically significant difference between the pairwise compared groups of animals. **Table 6** contains 37 compounds (42% of all compounds), the concentrations of which did not differ significantly between the compared groups.

Tables 1 and **2** are devoted to analyzing the amino acid composition. The first of them includes six amino acids that are also part of Actoflor®-S. The remaining 13 amino acids, coding and non-coding, are listed in **Table 2**. **Table 3** includes short-chain carboxylic acids, including three acids that are part of Actoflor®-S. Low-molecular carboxylic acids with additional functional hydroxy- and oxo-groups are listed in **Table 4**. **Table 5** contains the remaining identified compounds, including long-chain carboxylic acids, benzoic acid, etc.

4.1 Amino acids

The group of mice receiving Actoflor®-S, as expected, differed from the other groups by an increased content of almost all amino acids included in Actoflor®-S (**Table 1**), except for valine. From the fourth column of the table, it is evident that the maximum differences were observed for methionine, the average concentration of which in the blood was almost 13 times higher than in the control group. The glycine concentration increased by 5.5 times, and alanine and leucine—more than 2.5 times. In quantitative terms, as can be seen from **Table 1** (the penultimate line), the content of amino acids in the group receiving Actoflor®-S was about 4% of all identified compounds, and in the control group—only about 1.2%.

Other groups of experimental mice received biologically active additives that did not contain amino acids. However, in the "acetate" group, the total concentration of amino acids increased significantly (2.8 times) compared to the control, and the concentration of valine (11 times) and leucine (2.3 times) increased especially. Thus, acetate had a specific effect on the animals, increasing the synthesis of a number of amino acids by microbiota and, in particular, essential branched amino acids.

In contrast to acetate, the combined action of acetate and other carboxylic acids (**Table 1**, CAC group) resulted in a more than twofold decrease in the overall level of these amino acids in the blood and a significant decrease in the level of each of them when compared with the control group (alanine and leucine) or with the acetate (valine and leucine) and Actoflor (methionine and aspartic acid) groups.

In general, the data in **Table 1** indicate significant differences in the action of the dietary supplement Actoflor®-S and its components, that is, that the metabiotic Actoflor®-S has pronounced emergent properties.

The data in **Table 2** agree with the data in **Table 1** and generally confirm the conclusions made on its basis. As can be seen from this table, the effect of taking Actoflor®-S was not only an increase in the level of the amino acids contained in this dietary supplement but also of most of the others, so that their total percentage content in the metabolite pool increased more than 3-fold.

№	Metabolite (amino acid)	Concentration in control (C_c), %	Concentration relative to control (C/C_c)		
			Name of animal group		
			Actoflor	Acetate	CAC
1	L-Valine	0,084	0,85*	**11,29***#	0,55#
2	L-Alanine	0,553	**2,79***	1,41*	0,45*
3	Glycine	0,062	**5,51***	1,82*	0,32*
4	L-Leucine	0,262	**2,68**	**2,23**#	0,39*#
5	L-Aspartic acid	0,241	**4,43**	1,74	0,62*
6	L-Methionine	0,025	**12,95***	1,59*	0,74*
	Total conc., % of all	**1,227***	**4,05***	**2,883***	0,58*
	% of Control	100*	**330***	**235**#	48*#

Note: Significant ($p < 0,05$) differences from the control are highlighted in bold.
**significant ($p < 0,05$) differences between the Actoflor group and other groups, also marked with *.*
#significant ($p < 0.05$) differences between the Acetate and CAC groups.

Table 1.
Comparative analysis of the composition of amino acids (components of Actoflor®-S) in the blood of animals of the control group and groups receiving Actoflor®-S, or acetate, or CAC (carboxylic acid complex).

The content of non-coding α-aminoadipic acid (13.25-fold) and coding L-phenylalanine (8.4-fold) increased to the greatest extent. The level of oxyproline (non-coding amino acid) and L-serine (coding amino acid) increased more than 5-fold, the level of L-isoleucine increased almost 5-fold, and the level of threonine increased 3.45-fold. The level of non-coding L-pyroglutamic acid increased 1.7-fold, and the level of coding L-proline increased 2-fold. The latter generally had the highest concentration among all amino acids in the control group and the acetate and CAC groups. In the "Actoflor" group, its concentration reached 1.3% and was also higher than the concentration of all other amino acids not included in the composition of Actoflor®-S.

Among the four amino acids whose concentration in this group decreased, tyrosine is the most noteworthy. Its concentration in the control animals was one of the highest (0.436%), and under the action of Actoflor®-S, it fell more than 10 times. Tyrosine is an essential amino acid, but in the body of animals, it can be synthesized from phenylalanine, the concentration of which, as already noted, increased in this group more than 8 times (**Table 2**). Similarly, the concentration of glutamine and aminomalonic acid, which in the blood of animals of the control group were about 0.25 and 0.21%, respectively, decreased 5 times. Judging by the data obtained, the concentration of α-aminobutyric acid decreased more than 10 times, but this acid was

№	Metabolite (amino acid)	Concentration in control (C_c), %	Concentration relative to control (C/C_c)		
			Name of animal group		
			Actoflor	Acetate	CAC
1	L-Isoleucine	0,100	**4,82***	**3,24*#**	0,34*#
2	α-Aminobutyric acid	0,039	**0,09**	n.d.	n.d.
3	α-Methylalanine	0,009	2,77*	**0,21*#**	0,60#
4	L-Serine	0,127	**5,15***	**2,66*#**	0,61*#
5	L-Threonine	0,209	**3,45***	**2,21#**	**0,51*#**
6	L-Proline	0,634	**2,05***	1,54#	**0,45*#**
7	Oxyproline	0,060	**5,71***	1,44*	0,43*
8	Aminomalonic acid	0,214	0,20*	0,36#	**3,01*#**
9	L-Pyroglutamic acid	0,123	**1,73***	0,80	0,72*
10	α-Aminoadipic acid	0,056	**13,25***	2,98*	1,08*
11	L-Phenylalanine	0,056	**8,40***	2,93	0,73*
12	Glutamine	0,248	**0,20**	n.d.	n.d.
13	L-Tyrosine	0,436	**0,10**	n.d.	n.d.
	Total (13), conc., %	**2,311**	5,09	2,69	1,37
	Total (13), % of Control	100	220	117	59
	Total (10) conc., %	**1,598**	**4,99***	**2,69***	1,37*
	Total (10), % of Control	100	**314***	**170***	86*

Note: n.d.: no data; the remaining designations are the same as in ***Table 1****.*

Table 2.
Comparative analysis of the composition of amino acids not included in Actoflor®-S in the blood of animals of the control group and groups receiving Actoflor®-S, or acetate, or CAC (carboxylic acid complex).

DOI: http://dx.doi.org/10.5772/intechopen.1010019

found only in very small quantities, which does not allow us to consider this result completely reliable and indicative.

Regarding the nature of the differences from the control group, the "acetate" group was close to the "Actoflor" group. The concentration of 8 out of 10 amino acids common to these groups (**Table 2**) changed in a similar direction. The exceptions were α-methylalanine and L-pyroglutamic acid, the concentration of which in the "Actoflor" group was higher than in the control and lower in the "acetate" group. However, this difference from the control was not statistically confirmed for L-pyroglutamic acid.

In general, however, the changes in the amino acid composition in the "acetate" group, coinciding in direction with the "Actoflor" group, according to the data in **Table 2**, were expressed less than in the "Actoflor" group. This is evident both in the pairwise comparison of the data for individual amino acids and in the integral indicator "Total (10), % of Cont." in the last row of **Table 2**. The presence of similar changes between the "Actoflor" and "acetate" groups is also confirmed by the fact that there are only five reliable differences between the groups, whereas there are nine such differences between the Actoflor and CAC groups. The acetate and CAC groups differ noticeably less (6 differences) from each other.

4.2 Carboxylic acids

The analysis of the composition of carboxylic acids also showed quite unexpected results (**Table 3**). It was quite reasonable to expect that the mice that received the dietary supplements would have a high content of acetate and other acids in the blood, which are part of these dietary supplements. However, in all experimental groups, the concentration of acetate in the blood of animals was 1.4–2.2 times lower than in the control group. Between the experimental groups themselves, the concentration of acetate did not differ statistically. This made it possible to analyze the statistical significance not only in pairwise comparison of groups but also to calculate the P-value, combining all experimental groups together. As a result, it was confirmed that the acetate concentration in the combined group was significantly ($p = 0.003$) lower than in the control.

As for other acids, formic and succinic, their concentration was higher only in the blood of animals receiving Actoflor®-S, and in the CAC group, their concentration decreased rather than increased. The addition of acetate alone (acetate group), as expected, did not lead to a change in the level of these acids.

For acids not included in the dietary supplements studied, **Table 3** shows a significant decrease in the concentration of isobutyric, isovaleric, and a-valeric acids for all groups, with the greatest decrease occurring in the group receiving Actoflor®-S and the least in the CAC group. The decrease in the level of methylsuccinic acid was also significant for the two groups separately and for all three together ($p = 0.002$).

In addition to these acids, valeric acid was found in the blood of animals. Previously [7, 8], we showed that valeric acid is one of the few metabolites that have a noticeable beneficial effect on the condition of the intestinal epithelium and liver in the presence of their disorders, for example, in the case of inflammatory bowel disease or fatty liver disease. **Table 3** shows that the level of this acid in the blood increases only in the case of taking the complete dietary supplement Actoflor®-S, while its components did not increase this level and possibly even decrease it.

Speaking about **Table 3** as a whole (line 9), it can be noted that the CAC group was characterized by a statistically significant ($p = 0.007$) decrease in the level of

№	Metabolite (Acid)	Concentration in control (C_c), %	Concentration relative to control (C/C_c)		
			Name of animal group		
			Actoflor	Acetate	CAC
1	**Formic acid**	0,988	**3,43**	1,09	0,75[3]
2	**Acetic acid**	3661	0,62[1]	**0,46**	**0,73**[3]
3	**Succinic acid**	0,047	**3,07**	0,94	0,74[3]
4	Isobutyric acid	0,043	**0,20***	**0,34**	**0,61***
5	Isovaleric acid	0,043	**0,08***	**0,19**	**0,46***
6	a-Valeric acid	0,057	**0,10***	**0,33**	**0,49***
7	Valeric acid	0,021	**1,5**	1,06	0,72
8	Methylsuccinic acid	0,01	**0,03**	**0,07**	0,34[2]
9	Total conc., %	4,87	5,85	2,88	**3,53**
10	**% of Control**	100	120	59	72
11	Conc.(4 + 5 + 6 + 7 + 8),%	0,174	0,049	0,065	0,092
12	**% of Control**	100	28	37	53
13	Conc. (1 + 2 + 3),%	4,70	5,80	2,80	3,43
14	**% of Control**	100	124	60	73

[1]*p = 0.003 when combining all experimental groups.*
[2]*p = 0.002 when combining all experimental groups.*
[3]*p = 0.01 in a combined comparison of all three acids.*
Note: In the list of acids (lines No. 1–No. 8), the names of acids present in the dietary supplement are highlighted in bold. The remaining designations are the same as in ***Tables 1*** *and 2.*

Table 3.
Comparative analysis of the composition of low-molecular carboxylic acids (including components of Actoflor®-S) in the blood of animals of the control group and groups receiving Actoflor®-S, or acetate, or CAC (carboxylic acid complex).

absolutely all carboxylic acids. In the acetate group, the level of five out of eight carboxylic acids statistically significantly decreased, and the level of the remaining three acids (formic, succinic, and valeric) remained the same as in the control. The level of these three acids was even higher in the group receiving Actoflor®-S. In this case, the "control" level was significantly exceeded by 1.5–3 times. The increase in the level of these acids is probably not directly related to the presence of two of them (formic and succinic) in the composition of the taken dietary supplement Actoflor®-S, since the same acids and in the same amount were contained in the CAC supplement, and the result was the opposite. It is most likely that dietary supplements did not enter the blood directly but influenced the composition of the microbiota and thus changed the qualitative and quantitative composition of metabolites entering the host body from the microbiota.

If the first three acids in the table, which are part of the dietary supplement, are excluded from consideration, it turns out that the concentrations of the remaining acids in all experimental groups were significantly lower than in the control group (lines 11 and 12 in **Table 3**). This means that all the studied dietary supplements led to a decrease in the total level of carboxylic acids not included in them, and this was observed to the maximum extent for Actoflor®-S (28% of the control), to a lesser extent in the case of acetate (37%), and to and an even lesser extent in the CAC group (53%).

The total concentration of carboxylic acids (lines 9 and 10), as well as the total concentration of the first three carboxylic acids (lines 13 and 14) in the groups receiving acetate and CAC, were also lower and amounted to 59–60% and 72–73% of the control values, respectively. In general, this means that taking carboxylic acids (acetate and CAC) as a food additive (under the conditions of this experiment) led, oddly enough, to a decrease in their overall level in the blood of laboratory animals.

The most likely reason for all these changes is the same as above: the components of the dietary supplement entering the intestine have a regulatory effect on the composition of the microbiota, and the latter, in turn, affects the composition of low-molecular metabolites in the blood. However, a direct regulatory effect of the dietary supplement on the host organism cannot be ruled out. It is important to note that Actoflor®-S, in any case, behaves differently from its constituent parts—acetate and CAC. This indicates a coordinated action of all components of the drug and the presence of pronounced emergent properties both in Actoflor®-S itself and in its part—CAC.

Hydroxy- and oxocarboxylic acids play a significant role in the metabolic processes of the body. The most important and most represented hydroxyacid in the metabolome is lactic acid. For this reason, it was considered separately from other representatives of this class of acids. At the end of **Table 4**, a special section is devoted to it, from which it is evident that in the group of control mice, the concentration of lactic acid was almost 33% of the total number of metabolites identified. In the groups receiving acetate and CAC, this concentration was practically the same as in the control within the limits of statistical reliability. The concentration of lactic acid in the blood of animals is one of the most conservative indices of the metabolome, so it was surprising that in the group receiving Actoflor®-S, its concentration was significantly lower than in the control and the two experimental groups ($p = 0.05$). Reduced lactic acid levels may be at least one of the reasons for improved well-being and performance in athletes taking Actoflor®-S.

The concentration of most other metabolites (**Table 4**) decreased under the influence of all dietary supplements. The exception was 3-methyl-2-oxovaleric acid (No. 2 in the table), whose concentration significantly increased in the "Actoflor" group. In its structure, this acid is a ketone analog of isoleucine and is a product of its degradation.

In the acetate group, the concentration of α-hydroxybutyric acid (No. 3 in the table) increased on average by 2 times, but due to the large spread of its concentrations within the group, this increase was statistically insignificant compared to the control; however, when compared with the CAC group, the significance of the difference was greater and corresponded to $p = 0.05$.

The only compound whose concentration significantly decreased in all three groups was glyceric acid (No. 7 in the table), while the acetate and Actoflor groups significantly ($p = 0.026$ and $p = 0.004$, respectively) differed in lower concentrations from the CAC group but did not differ from each other. Unfortunately, we do not have data on the concentrations of 2,3-dihydroxybutyric (No. 8), β-hydroxycaproic (No. 6), and 3-methyl-2-oxovaleric (No. 2) acids in mice receiving acetate or CAC. In the Actoflor group, the concentration of these compounds was significantly lower than in the control. Perhaps, by analogy with glyceric acid, their concentrations similarly decreased in these groups.

№	Metabolite (Acid)	Concentration in control (C_c), %	Concentration relative to control (C/C_c)		
			Name of animal group		
			Actoflor	Acetate	CAC
1	Glycolic	0,498	0,83*	**0,39***	0,75
2	3-Methyl-2-oxovaleric	0,046	1,50	n.d.	n.d.
3	α-Hydroxybutyric	0,258	0,74	2,07#	0,40#
4	β-Hydroxypropionic	0,064	0,58*	**0,02*#**	**0,43#**
5	2-Methyl-2-hydroxybutyric	0,064	1,08*	**0,25***	0,7
6	β-Hydroxycaproic	0,016	**0,5**	n.d.	n.d.
7	Glyceric acid	0,6	**0,21***	**0,35#**	**0,71*#**
8	2,3-dihydroxybutyric	0,058	**0,21**	n.d.	n.d.
9	Hydroxymalonic	0,022	**0,20**	**0,17**	0,71
10	2,4-Dihydroxybutyric	0,039	0,73	0,31#	1,08#
11	3-Hydroxy-3-methylglutaric	0,097	**0,20***	**0,01#**	1,26*#
12	l-Threonic	0,319	**0,62**	1,12	1,05
13	Citric	0,329	1,29*	0,45*#	1,43#
14	Total conc., %	2,41	1,67	1,51	1,96
14	% of Control	100	69,5	62,6	81,3
16	Sum without No.2, 6 and 8	2,29	1,58	1,51	1,96
18	% of Control	100	69,2	65,8	85,6
		Lactic acid			
19	C/C_c.	1	0,74*	1,22*	0,97*
20	Concentration, % of total	32,8	24,3*	40,1*	31,9*
21	% of Control	100	74*	122	97

Note: All designations are the same as in the tables above.

Table 4.
Comparative analysis of the composition of short-chain hydroxy- and oxocarboxylic acids in the blood of animals of the control group and groups receiving Actoflor®-S, or acetate, or CAC (carboxylic acid complex).

The concentration of β-hydroxypropionic acid (No. 4), a structural isomer of lactic acid, did not differ significantly from the control in the Actoflor group but was significantly lower than the control in the CAC and acetate groups. The latter was characterized by the lowest level of β-hydroxypropionic acid and statistically significantly differed not only from the control group but even from the Actoflor ($p = 0.03$) and CAC ($p = 0.004$) groups. The concentrations of glycolic acid (No. 1) and 2-methyl-2-hydroxybutyric acid (No. 5) were also significantly lower in the acetate group compared to the control and Actoflor groups (**Table 4**).

Among the compounds listed in **Table 4**, there are two dicarboxylic hydroxy acids: hydroxymalonic acid (No. 9) and 3-hydroxy-3-methylglutaric acid (No. 11). The concentration of both acids was statistically significantly lower than the control in the Actoflor and acetate groups (**Table 4**). The content of 3-hydroxy-3-methylglutaric acid also significantly differed between the Actoflor/CAC ($p = 0.009$)

and CAC/acetate (p = 0.002) groups; no such intergroup difference was found for hydroxymalonic acid.

The concentration of 2,4-dihydroxybutyric acid (No. 10) did not differ significantly from the control in any of the groups. However, one significant (p = 0.05) difference was still found for it—between the CAC and acetate groups.

The concentrations of the last two acids (No. 12 and No. 13), l-threonic and citric, were significantly higher than most other acids in the table. The concentration of the first of them in the Actoflor group was statistically significantly lower than in the control (**Table 4**), and the concentration of the second did not differ significantly from the control in any of the groups but differed between the groups receiving Actoflor®-S and acetate (p = 0.03) and acetate and CAC (p = 0.05).

The following **Table 5** lists the metabolites that were not included in any of the previous tables but the concentrations of which were statistically significantly different from the control and/or between the groups. Unlike the previous tables, the main reliable differences from the control group were associated with the acetate group (seven out of nine possible); only two differences fell to the share of the Actoflor group and only one difference to the CAC group.

The Actoflor group differed from the acetate group in three positions: uracil (p = 0.05), creatinine (p = 0.009), and tetradecanoic acid (p = 0.008). It differed from the CAC group only in the concentration of dodecanoic acid (p = 0.004).

The greatest number of differences (6) were found when comparing the CAC and acetate groups: benzoic acid (p = 0.004), monophosphate (p = 0.022), nonanoic acid (p = 0.017), creatinine (p = 0.01), tetradecanoic acid (p = 0.05), and cis-6-octadecenoic acid (p = 0.004).

One of the substances listed in **Table 5** remained unnamed, but it was characterized by retention time and characteristic mass spectrum. These characteristics will allow identifying this substance in the future.

№	Metabolite	Concentration in control (C_c), %	Concentration relative to control (C/C_c)		
			Name of animal group		
			Actoflor	Acetate	CAC
1	Benzoic acid	0,135	1,54	**0,02**#	2,65#
2	Monophosphate	0,544	**1,66**	**3,43**#	0,97#
3	Uracil	0,047	0,82*	**0,17***	0,68
4	Nonanoic acid	0,033	0,80	**0,04**#	0,46#
5	Creatinine	0,066	1,53*	**0,14***#	0,88#
6	Not identified	0,144	**0,24**	n.d.	n.d.
7	Dodecanoic acid	0,052	0,40*	0,28	**0,08***
8	Tetradecanoic acid	0,076	2,60*	**0,16***#	1,31#
9	cis-6-octadecenoic	0,123	0,92	**0,07**#	1,20#

Note: All designations are the same as in the ***Tables 1–3****.*

Table 5.
Comparative analysis of the composition of other metabolites in the blood of animals of the control group and groups receiving Actoflor®-S, or acetate, or CAC (carboxylic acid complex).

№	Metabolite	№	Metabolite
1	Butyric acid	20	Malic acid
2	Oxalic acid	21	Undecanoic acid
3	Hydrogen sulfide	22	Norleucine
4	N,N-Dimethylglycine	23	Maleic acid
5	Ethylene glycol	24	3-Hydroxyphenylpropionic acid
6	α-Hydroxyisobutyric acid	25	Glycerophosphate
7	Pyruvic acid	26	Nonanedioic acid
8	β-Hydroxybutyric acid	27	n-Pentadecanoic acid
9	α-Hydroxyvaleric acid	28	cis-9-Hexadecenoic acid
10	Urea	29	Hexadecanoic acid
11	β-Hydroxyvaleric acid	30	9,12-Octadecadienoic acid (Z,Z)-
12	Acetoacetic acid	31	trans-9-Octadecenoic acid
13	Octanoic acid	32	trans-13-Octadecenoic acid
14	Glycerol	33	trans-11-Octadecenoic acid
15	Fumaric acid	34	Octadecanoic acid
16	2-Methylbenzoic acid	35	Arachidonic acid
17	α-phenylpropionic acid	36	cis-4,7,10,13,16,19-Docosahexaenoic acid
18	3,4-dihydroxybutyric acid	37	Cholesterol
19	Decanoic acid	38	Ethanolamine

Table 6.
List of compounds for which statistically significant differences in concentrations between animal groups were not detected.

The final **Table 6** contains compounds whose concentrations do not differ statistically significantly between the groups. These substances include such metabolically significant compounds as pyruvic, acetoacetic, fumaric, malic, and oxalic, as well as butyric and hydroxybutyric acids and their isomers; various saturated and unsaturated medium- and long-chain acids, including those of microbial origin, glycerol, glycerophosphate, urea, cholesterol, hydrogen sulfide, and some other compounds. The homeostasis of these substances is apparently too important for the macroorganism and therefore is not subject to (or is subject to a lesser degree) regulation by metabolite preparations.

5. Therapeutic potential of Actoflor®-S in the prevention and treatment of intestinal, liver, and metabolic disorders

The therapeutic potential of Actoflor®-S is determined, first, by its effect on the composition and physiological functions of the microbiota, which results in a change in the qualitative and quantitative composition of the flow of metabolites entering the host organism. In the host's blood, metabolites have a regulatory effect on all its organs and tissues, helping maintain health or, on the contrary, causing changes that favor the development of a particular pathology. Except for potent poisons, hormones, neurotransmitters, and some other similar molecules, the host organism does not react to a single substance but to a certain set of different molecules, as was clearly shown in this study using the example of Actoflor®-S dietary supplement and its components. The effect of a complete dietary supplement differed significantly from the effect of its parts; in other words, the drug had pronounced emergent properties. For example, in animals receiving Actoflor®-S, compared to other groups of animals, the concentration of almost all amino acids in the blood significantly increased, while the concentration of carboxylic acids and their derivatives decreased. In the group receiving only acetate, the concentration of amino acids also tended to increase, but this increase was less. And finally, in the CAC group, the concentration of amino acids was noticeably lower than in the control and, especially, in the other groups. Since the host itself could not synthesize essential amino acids, it is logical to conclude that their synthesis was carried out by the microbiota and was stimulated by the amino acids from Actoflor®-S. Acetate also stimulated the synthesis of amino acids but exhibited this ability to a significantly lesser extent than the amino acids themselves.

The latter property of acetate is not surprising, since it is a desirable substrate for a number of intestinal bacteria and the starting product in many biosynthetic processes. What was surprising was that in combination with three other carboxylic acids (i.e., as part of the CAC), their ability to stimulate the synthesis of amino acids completely disappeared. Actoflor®-S, which is the same CAC but with the addition of several amino acids, was again able to stimulate their synthesis, including the synthesis of amino acids not included in Actoflor®-S.

One of the exceptions to the general rule was the abnormal "behavior" of valine. Its concentration, unlike other amino acids, did not increase against the background of taking Actoflor®-S but increased more than 10 times in animals receiving acetate. In other words, adding any other bacterial metabolites to acetate (i.e., carboxylic acids and other components of Actoflor®-S) led to acetate losing its ability to induce valine synthesis.

For the CAC group, a similar exception was aminomalonic acid, the concentration of which increased only under the influence of CAC, while other metabolites suppressed its synthesis by 3–5 times. Since aminomalonic acid is metabolically associated with serine and threonine, the concentration of which increased under the influence of acetate, it can be assumed that other CAC acids contributed to the degradation of these amino acids with the formation of aminomalonic acid. In any case, when comparing all three groups of animals, the concentration of aminomalonic acid negatively correlated with the concentrations of serine and threonine. Aminoadipic acid, whose concentration, like other amino acids, was highest in the Actoflor group, is a metabolite in the biochemical pathway of degradation (and synthesis) of lysine. This suggests that its high concentration is explained, at least in part, by the fact that its precursor lysine is included in Actoflor®-S.

The effect of all the studied additives on the level of carboxylic acids, as well as their hydroxy- and oxo-derivatives, was similar—their concentration decreased or, at least, remained at the control level. The main exceptions were associated with the components of Actoflor®-S and were observed in the corresponding group of animals (**Table 3**). The concentration of formic and acetic acids in this group was approximately 3 times higher than the control level. The concentration of all other acids (**Table 3**, line 12) was only 28% of the control level. However, the CAC group contained the same acids as Actoflor®-S, but, oddly enough, their concentration in the blood remained below the control group.

Earlier, [9] we studied the metabolic profile of blood in patients with ulcerative colitis (UC) and showed that increased or decreased concentrations of a number of compounds from **Tables 1–5** are observed in UC patients. For example, all women with UC had a decreased level of succinic acid, and most of them also had a decreased level of l-alanine, l-aspartic acid, l-threonine, l-proline, α-aminoadipic acid, l-phenylalanine, glyceric acid, and creatinine. Since the level of all these compounds increased under the influence of Actoflor®-S, it can be assumed that Actoflor®-S has the corresponding therapeutic potential in treating this disease. Similarly, in patients with UC, the level of α-hydroxybutyric and 2,3-dihydroxybutyric acids was increased, while under the influence of Actoflor®-S, this level decreased.

As can be seen from the data obtained, the property to normalize the composition of the metabolome is inherent only to the entire set of components of Actoflor®-S but is not characteristic of its components—neither acetate nor the entire CAC, which accounts for 85% of the dry weight of the drug. The results of the work also indicate that all metabolites of the metabolome are important participants in maintaining health, often having individual activity, but to a greater extent, their activity is manifested through combined action.

An example of the synergistic effect of metabolites is the effect of valeric acid (valerate) in combination with other metabolites. **Table 3** shows that the concentration of this acid, practically the only one among the acids if we do not consider the components of Actoflor®-S, significantly increased in the metabolome of animals. This acid itself had a positive effect on intestinal epithelial cells but acted to a much greater extent in combination with other metabolites of microbiota bacteria and mucin components supplied by the host organism [7]. In microbiota, valerate producers are intestinal bacteria such as *Megasphaera elsdenii*, *Megasphaera massiliensis*, *Oscillibacter valericigenes*, etc. [10]. In relevant studies, valeric acid levels were shown to be negatively associated with the severity of ulcerative colitis, monocyte population, and proinflammatory IL-6 levels [11]. Fecal levels of valeric and isovaleric acids were negatively correlated with the severity of Crohn's disease in children [12]. Valerate is currently considered a promising therapeutic and prophylactic agent for both inflammatory bowel diseases and metabolic disorders [11, 13]. The decrease in lactic acid levels with Actoflor®-S may be because certain microbiota bacteria increase the rate of its consumption under the influence of Actoflor®-S. In the context of inflammatory bowel diseases and metabolic disorders, this reduction can be considered as a potential therapeutic and/or preventive effect [14–17]. In addition, reducing lactic acid levels, as already noted, is very attractive for athletes, since it leads to an increase in their endurance during prolonged physical exertion [18].

The reduction of methylsuccinate levels by Actoflor®-S may be beneficial in both metabolic disorders and intestinal inflammation. The level of methylsuccinic acid in urine is increased in children with IBD [19, 20]. In addition, an increase in methylsuccinate levels is characteristic of type 2 diabetes mellitus and may reflect more intense catabolism of isoleucine, the dysregulation of which is associated with a high risk of developing diabetes and insulin resistance [21]. Discussing the changes in the level of amino acids not included in the drug revealed under the influence of Actoflor®-S, it is worth noting that in metabolic disorders such as non-alcoholic fatty liver disease (NAFLD), a violation of the metabolism of aromatic amino acids and branched-chain amino acids has pathogenetic significance and may be associated with a change in the taxonomic composition and functional activity of the intestinal microbiota [22].

An attractive effect of Actoflor®-S is the increase in the level of phenylalanine. Phenylalanine, a significant part of which is produced by intestinal bacteria such as *Bacteroides thetaiotaomicron*, *Escherichia* spp., *Akkermansia muciniphila*, *Subdoligranulum variabile,* and *Intestinibacter bartlettii*, is capable of exerting antioxidant and anti-inflammatory effects. In addition, it regulates the release of intestinal hormones and glucose tolerance, reducing the risk of obesity and diabetes [22]. Impaired microbial production of phenylalanine in intestinal dysbiosis can contribute to the pathogenesis of NAFLD and other metabolic disorders.

However, tyrosine metabolism is most impaired in patients with NAFLD, especially in children and adolescents, and is also associated with the risk of developing hyperglycemia, insulin resistance, metabolic syndrome, and diabetes. Blood tyrosine levels positively correlate with the severity of liver steatosis [23]. Excess tyrosine, when its microbial metabolism is impaired, is degraded to acetyl-CoA through ketogenesis, stimulating the synthesis of fatty acids and promoting lipid deposition in the liver [22]. Since blood tyrosine levels are positively associated with NAFLD, insulin resistance, diabetes, and obesity [24, 25], we believe that a decrease in its serum level during the use of Actoflor®-S can be considered as a positive (therapeutic and/or preventive) effect in the context of metabolic disorders.

We assessed the almost 5-fold increase in the level of L-isoleucine revealed during the use of Actoflor®-S as metabolically beneficial since the addition of isoleucine in experimental studies contributed to a decrease in blood glucose and body fat levels and also led to health-related changes in the intestinal microbiota [26]. The more than 3-fold increase in blood L-threonine concentration in the Actoflor®-S group can also be regarded as a positive effect since it has been previously shown that threonine deficiency enhances mitochondrial uncoupling in the liver, causing triglyceride accumulation in the liver, and threonine supplementation can prevent the development of hyperlipidemia and enhance lipid metabolism in the liver [23].

6. Conclusions

The Actoflor®-S preparation was developed as a synthetic analog of the Actoflor preparation, which contained highly active "bios" of the probiotic bacterium *E. coli* M-17 obtained under special conditions [5]. Naturally, it could not contain all the components of natural preparation. The synthetic design was based on two main principles: the principle of reasonable diversity of components (according to this principle, the number of components was limited) and the principle of reliability, implying partial duplication of the functions of one component by another. These principles allow varying the preparation composition within certain limits while maintaining all its main functions. Taking these principles into account, the manufacturer of the preparation, Grotex LLC, is constantly working on its improvement and adaptation for certain groups of the population. In particular, taking into account the greater sensitivity of children to carboxylic acids, Actoflor® Kids was developed, containing fewer carboxylic acids and more amino acids and vitamins. According to the results obtained in this study, "strengthening" the amino acid part of the drug should contribute to an increase in the total pool in the blood of children, suppression of undesirable inflammatory reactions, and a general healing effect due to strengthening the communication links of the child's body with its microbiota.

Acknowledgements

The authors express their deep gratitude to V.A. Utsal for assistance in conducting chromatographic analysis and to T.S. Egorova for assistance in working with laboratory animals.

Author details

Timur Vakhitov[1*] and Stanislav Sitkin[1,2]

1 Institute of Experimental Medicine, Saint Petersburg, Russian Federation

2 Almazov National Medical Research Centre, Saint Petersburg, Russian Federation

*Address all correspondence to: tim-vakhitov@yandex.ru

References

[1] Pasteur L. Mémoires sur la fermentation alcoolique. Paris: Impriiuerie de Mallet-Bacueliur, rue du Jardinet, 12; 1860. 106 p

[2] Wildiers E. Nouvelle substance indispensable au developpement de la levure. La Cellule. 1901;**18**:313-333

[3] Meisel MN. Functional Morphology of Yeast Organisms. Moscow: Publishing House of the USSR Academy of Sciences; 1950. 368 p

[4] Rahn O. Ueber den Einfluss der Stoffwechselprodukte auf das Wachstum der Bakterien. Zentr. Bakt. Parasitenk. 1906;**II**,**16**:417-429 & 609-617

[5] Vakhitov TY. Principles of Metabolic Regulation and Communication in Bacteria. M: LENAND; 2019. 500 p. (In Russian). ISBN: 978-5-9710-6426-8

[6] Weichart DH, Kell DB. Characterization of an autostimulatory substance produced by *Escherichia coli*. Microbiology (Reading). 2001;**147**(Pt. 7):1875-1885. DOI: 10.1099/00221287-147-7-1875

[7] Sall T, Sitkin S, Lazebnik L, Vakhitov T. Effects of gut microbiota metabolites on the intestinal epithelial cell viability, barrier function, IL-8 secretion, and triglyceride accumulation in cell models of IBD and NAFLD. European Journal of Case Reports in Internal Medicine (EJCRIM). 2023;**10**(Suppl. 1):222. DOI: 10.12890/2023_V10Sup1

[8] Sall T, Sitkin S, Lazebnik L, Vakhitov T. Postbiotic metabolites as intestinal barrier protectors during *Helicobacter pylori* eradication therapy. Microbiota in Health and Disease. 2023;**5**:e857(155). DOI: 10.26355/mhd_20239_857

[9] Vakhitov TY, Kononova SV, Demyanova EV, Morugina AS, Utsal VA, Sall TS, et al. Serum metabolomic profile in patients with ulcerative colitis: Pathophysiological role, diagnostic and therapeutic implications. Voprosy Detskoj Dietologii (Pediatric Nutrition). 2023;**21**(5):5-15. (In Russian). DOI: 10.20953/1727-5784-2023-5-5-15

[10] Yuille S, Reichardt N, Panda S, Dunbar H, Mulder IE. Human gut bacteria as potent class I histone deacetylase inhibitors in vitro through production of butyric acid and valeric acid. PLoS One. 2018;**13**(7):e0201073. DOI: 10.1371/journal.pone.0201073

[11] Liu M, Zhang Y, Liu J, Xiang C, Lu Q, Lu H, et al. Revisiting the role of valeric acid in manipulating ulcerative colitis. Inflammatory Bowel Diseases. 2024;**30**(4):617-628. DOI: 10.1093/ibd/izad187

[12] Wang Y, Gao X, Zhang X, Xiao F, Hu H, Li X, et al. Microbial and metabolic features associated with outcome of infliximab therapy in pediatric Crohn's disease. Gut Microbes. 2021;**13**(1):1-18. DOI: 10.1080/19490976.2020.1865708

[13] Kumari B, Kumari U, Singh DK, Husain GM, Patel DK, Shakya A, et al. Molecular targets of valeric acid: A bioactive natural product for endocrine, metabolic, and immunological disorders. Endocrine, Metabolic & Immune Disorders Drug Targets. 2024;**24**(13):1506-1517. DOI: 10.2174/0118715303262653231120043819

[14] Singhal R, Shah YM. Oxygen battle in the gut: Hypoxia and

hypoxia-inducible factors in metabolic and inflammatory responses in the intestine. The Journal of Biological Chemistry. 2020;**295**(30):10493-10505. DOI: 10.1074/jbc.REV120.011188

[15] Taylor SJ, Winter MG, Gillis CC, Silva LAD, Dobbins AL, Muramatsu MK, et al. Colonocyte-derived lactate promotes *E. coli* fitness in the context of inflammation-associated gut microbiota dysbiosis. Microbiome. 2022;**10**(1):200. DOI: 10.1186/s40168-022-01389-7

[16] Ahola-Olli AV, Mustelin L, Kalimeri M, Kettunen J, Jokelainen J, Auvinen J, et al. Circulating metabolites and the risk of type 2 diabetes: A prospective study of 11,896 young adults from four Finnish cohorts. Diabetologia. 2019;**62**(12):2298-2309. DOI: 10.1007/s00125-019-05001-w

[17] Hameed A, Mojsak P, Buczynska A, Suleria HAR, Kretowski A, Ciborowski M. Altered metabolome of lipids and amino acids species: A source of early signature biomarkers of T2DM. Journal of Clinical Medicine. 2020;**9**(7):2257. DOI: 10.3390/jcm9072257

[18] Hoff J, Støren Ø, Finstad A, Wang E, Helgerud J. Increased blood lactate level deteriorates running economy in world class endurance athletes. Journal of Strength and Conditioning Research. 2016;**30**(5):1373-1378. DOI: 10.1519/JSC.0000000000001349

[19] Martin FP, Su MM, Xie GX, Guiraud SP, Kussmann M, Godin JP, et al. Urinary metabolic insights into host-gut microbial interactions in healthy and IBD children. World Journal of Gastroenterology. 2017;**23**(20):3643-3654. DOI: 10.3748/wjg.v23.i20.3643

[20] Filimoniuk A, Daniluk U, Samczuk P, Wasilewska N, Jakimiec P, Kucharska M, et al. Metabolomic profiling in children with inflammatory bowel disease. Advances in Medical Sciences. 2020;**65**(1):65-70. DOI: 10.1016/j.advms.2019.12.009

[21] Urpi-Sarda M, Almanza-Aguilera E, Llorach R, Vázquez-Fresno R, Estruch R, Corella D, et al. Non-targeted metabolomic biomarkers and metabotypes of type 2 diabetes: A cross-sectional study of PREDIMED trial participants. Diabetes & Metabolism. 2019;**45**(2):167-174. DOI: 10.1016/j.diabet.2018.02.006

[22] Shcherbakova ES, Sall TS, Sitkin SI, Vakhitov TY, Demyanova EV. The role of bacterial metabolites derived from aromatic amino acids in non-alcoholic fatty liver disease. Almanac of Clinical Medicine. 2020;**48**(6):375-386. (In Russian). DOI: 10.18786/2072-0505-2020-48-066

[23] Deng Y, Hu M, Huang S, Fu N. Molecular mechanism and therapeutic significance of essential amino acids in metabolically associated fatty liver disease. The Journal of Nutritional Biochemistry. 2024;**126**:109581. DOI: 10.1016/j.jnutbio.2024.109581

[24] Chen T, Zheng X, Ma X, Bao Y, Ni Y, Hu C, et al. Tryptophan predicts the risk for future type 2 diabetes. PLoS One. 2016;**11**(9):e0162192. DOI: 10.1371/journal.pone.0162192

[25] Arjmand B, Ebrahimi Fana S, Ghasemi E, Kazemi A, Ghodssi-Ghassemabadi R, Dehghanbanadaki H, et al. Metabolic signatures of insulin resistance in non-diabetic individuals. BMC Endocrine Disorders. 2022;**22**(1):212. DOI: 10.1186/s12902-022-01130-3

[26] Crovesy L, El-Bacha T, Rosado EL. Modulation of the gut microbiota by

probiotics and symbiotics is associated with changes in serum metabolite profile related to a decrease in inflammation and overall benefits to metabolic health: A double-blind randomized controlled clinical trial in women with obesity. Food & Function. 2021;**12**(5):2161-2170. DOI: 10.1039/d0fo02748k

Chapter 8

Potential Probiotics for the Therapy of Metabolic Dysfunction-Associated Steatotic Liver Disease

Xiayun Li, Liyan Yu, Zonggen Peng, Tingxia Lv and Hu Li

Abstract

Metabolic dysfunction-associated steatotic liver disease (MASLD), a prevalent metabolic disorder globally, has presented an urgent need for effective therapeutic agents. The complex pathogenesis of MASLD and the failure of numerous clinical trials targeting this disease have prompted the exploration of probiotic intervention as a novel therapeutic strategy. Emerging evidence underscores the pivotal role of the gut-liver axis in MASLD progression, particularly through dysregulation of gut microbiota composition, impaired intestinal barrier integrity, and aberrant bacterial metabolite signaling. Preclinical studies indicate that specific probiotic strains may ameliorate MASLD by restoring microbial homeostasis, fortifying gut barrier function, and attenuating hepatic inflammation via gut-liver crosstalk. Preliminary clinical trials further support the beneficial effects of probiotics in reducing hepatic steatosis and improving metabolic parameters. Future research should prioritize elucidating strain-specific mechanisms, optimizing probiotic formulations, and addressing challenges related to long-term efficacy, safety, and personalized therapeutic regimens. This review comprehensively evaluates current evidence on probiotic applications for MASLD treatment and highlights critical directions for advancing translational research in this field.

Keywords: MASLD, probiotics, gut-liver axis, gut microbiota, metabolite signaling

1. Introduction

Metabolic dysfunction-associated steatotic liver disease (MASLD), previously known as nonalcoholic fatty liver disease (NAFLD), is characterized by abnormal lipid accumulation in the liver and is often associated with metabolic risks such as type 2 diabetes mellitus, obesity, and other metabolic disorders [1]. The current prevalence of MASLD in adults is approximately 25–30% and has increased rapidly, thereby contributing to the global burden. Despite decades of research and hundreds of clinical trials aimed at exploring therapeutic targets and drugs, progress has been hindered by the complex pathogenesis of MASLD. Although the first drug, resmetirom, has received approval for treating metabolic dysfunction-associated steatohepatitis (MASH), the progressive subset of MASLD, in 2024, there remains an urgent need for novel treatment strategies [2]. Given the established connections between gut microbiota dysbiosis, gut barrier dysfunction, and alterations in gut bacterial metabolites with MASLD, as well as the increasingly documented beneficial effects of gut microbes in alleviating metabolic disorders, probiotic interventions have emerged as a promising therapeutic approach [3].

The human gastrointestinal tract harbors approximately 100 trillion bacteria, and the gut microbiota is often referred to as the "second genome" of the human body. It is crucial to promote digestion and nutrient absorption, influence host metabolism, and participate in immune and drug metabolism processes. The colon, characterized by its anaerobic environment and largest biomass, is predominantly populated by anaerobic bacteria. These bacteria are distributed across several phyla, including *Bacillota* (*Firmicutes*), *Bacteroidota* (*Bacteroidetes*), *Actinomycetota* (*Actinobacteria*), *Pseudomonadota* (*Proteobacteria*), and *Verrucomicrobiota* (*Verrucomicrobia*). Additionally, the gut hosts a significant presence of viruses, fungi, and archaea [4]. With advancements in metagenomics, metabolomics, lipidomics, and transcriptomics, the complex interactions between the host and various microorganisms are progressively elucidated. Currently, gut dysbiosis is linked to numerous diseases, such as obesity, type 2 diabetes, hepatic steatosis, gastrointestinal disorders, and cancer. Consequently, gut microbiota has emerged as a novel target for preventing and treating metabolic diseases. This article investigates the role of gut microbiota in the pathogenesis of MASLD and explores probiotic therapeutic strategies.

2. The association between the gut-liver axis and MASLD

The bile secreted by the liver is released into the small intestine via the biliary system, where it participates in digestion and absorption. Concurrently, gut components and microbiota metabolites are first exposed to the liver through the portal vein. Based on the developmental and anatomical relationship between the liver and the gut, Marshall first proposed the concept of the gut-liver axis in 1998 [5]. The gut microbiota serves as a bridge connecting the liver, gut, and portal vein, acting as both an inducer and a potential therapeutic target for the onset of liver and other metabolic diseases. Within the gut-liver axis, gut microbiota, gut barrier, and the presence of gut bacterial metabolites contribute to the progression or treatment of MASLD.

2.1 Dysbiosis of the gut microbiota

Dysbiosis of the gut microbiota is observed at every stage of metabolic disease. Studies comparing the gut microbiota composition of patients at various stages of NAFLD through 16S rRNA gene sequencing analysis have demonstrated that the α-diversity is highest in the control group, followed by individuals with obesity, nonalcoholic steatohepatitis (NASH), and those with NAFLD [6]. At the phylum level, patients with fatty liver or steatohepatitis exhibit significantly increased abundances of *Pseudomonadota* and *Bacillota*, alongside a decrease in the abundance of *Bacteroidota* [7]. This trend reflects an increase in the *Bacillota* (*Firmicutes*) to *Bacteroidota* (*Bacteroidetes*) (F/B) ratio [8]. Similar alterations in the gut microbiome of children with NAFLD, NASH, and obesity compared to healthy controls were also reported in a previous study [9]. NAFLD patients displayed higher abundances of

Anaerococcus, *Bradyrhizobium*, *Dorea*, *Peptoniphilus*, *Propionibacterium acnes*, and *Ruminococcus*, while *Oscillospira* and *Rikenellaceae* were found to be less abundant. Changes in microbial abundance suggest that these bacteria may function as either beneficial or pathogenic microbes during the progression of NAFLD, providing a foundation for microbiota-targeted research. For instance, liraglutide treatment has been shown to modulate gut microbiota by reducing the abundance of *Pseudomonadota* in two obesity animal models, which is associated with an increased abundance of *Akkermansia* [10]. These findings indicate that gut microbiota and the development of NAFLD are interrelated, and targeting the microbiota presents a promising therapeutic strategy.

2.2 Gut barrier damage

When the intestinal epithelial barrier (IEB) is compromised, gut inflammation leads to damage to the gut vascular barrier (GVB), allowing intestinal components to enter the liver through the portal vein [11]. The liver is the first organ to encounter gut-derived bacteria and their pathogen-associated molecular patterns (PAMPs). These metabolism-associated molecular patterns (MAMPs) or PAMPs can be recognized by the liver's innate immune cells, triggering and maintaining an inflammatory response, thereby promoting the development of NAFLD [12]. Compared to patients without NAFLD, NAFLD patients exhibited 38–40% higher levels of lipopolysaccharide (LPS) in serum [13]. When LPS activates the signaling pathway through Toll-like receptor 4 (TLR4), it triggers the activation of nuclear factor-κB (NF-κB), which in turn activates inflammasomes. Additionally, endotoxins can directly damage hepatocytes, activate Kupffer cells to produce inflammatory cytokines, and release reactive oxygen species (ROS), leading to inflammation, steatohepatitis, and liver fibrosis [14].

2.3 Gut bacterial metabolites

2.3.1 Short-chain fatty acids

Acetate, propionate, and butyrate are the most prevalent short-chain fatty acids (SCFAs) that are produced by gut microbiota, which could regulate immune responses through G-protein-coupled receptors (GPCRs) such as GPR43, GPR41, and GPR109A in immune and intestinal epithelial cells [15, 16]. Key SCFA-producing bacteria include *Bifidobacterium*, *Lactobacillus*, *Lachnospiraceae*, and *Ruminococcaceae* [17]. Other bacteria, such as *Akkermansia muciniphila*, can convert dietary fiber into SCFAs, thereby contributing to reversing antibiotic or high-fat diet-induced metabolic dysregulation [18, 19]. It is also reported that *Lactobacillus sakei* MJM60958 alters SCFAs production in the cecum by reducing lactate and increasing acetate production [20]. Conversely, SCFAs in the gut also regulate the composition of gut microbiota, with studies showing that supplementation with acetate or propionate in high-fat diet-induced obese mice significantly reduced the gut F/B ratio [21].

2.3.2 Bile acids

A small portion of bile acids (BAs) escapes reabsorption in the gastrointestinal tract and undergoes various biotransformations, such as oxidation, esterification, hydroxyl-shifting isomerization, and desulfation, mediated by gut bacteria [22]. This process generates secondary bile acids with enzymatic activity in many bacteria [23, 24].

The gut microbiota modifies BAs to produce conjugated bile acids (CBA), which return to the liver and promote hepatic steatosis by downregulating the acetylation of H3K9, H3K14, H3K18, and H3K27 [25]. In intestinal epithelial cells, the activation of the farnesoid X receptor (FXR) induces FGF19/15, which subsequently activates fibroblast growth factor receptor 4 (FGFR4) in hepatocytes. This cascade leads to the upregulation of SHP and the downregulation of CYP7A1 expression, thereby reducing bile acid levels and altering bile acid composition to influence lipid absorption in the gut [26].

2.3.3 Endogenous ethanol and endotoxins

Endotoxins and other harmful microbial metabolites can enter the portal vein through a compromised gut barrier, leading to an upregulation of pro-inflammatory cytokine production in the liver, exacerbating NAFLD development [27]. Recently reported gram-positive bacteria, including *Clostridium asparagiforme*, *Clostridium bolteae*, *Clostridium clostridioforme*, and *Clostridium citroniae*, have been shown to increase gut permeability, thereby facilitating the translocation of bacterial metabolites and cellular components from the gut to the liver [28]. This process triggers further inflammation and promotes the progression of NAFLD. In addition to bacteria, the gut yeast Pichia pastoris is also linked to the pathophysiology of NASH through fructose-dependent endogenous ethanol and triglyceride production in a species- and strain-specific manner [29].

2.3.4 Choline

When the gut microbiota is dysregulated, it metabolizes choline into trimethylamine (TMA) and subsequently oxidizes into trimethylamine N-oxide (TMAO) in the liver, which enters the bloodstream and promotes inflammation and insulin resistance. TMAO also influences lipid absorption and cholesterol homeostasis by converting cholesterol into bile acids, thereby increasing the risk of NAFLD [30]. A previous study demonstrated that the improvement of NAFLD in methionine- and choline-deficient diet-induced mice using fibroblast growth factor 21 (FGF21) is mediated by the gut microbiota and choline [31]. These findings suggest that restoring the gut microbiota may serve as a viable pathway for regulating abnormal choline metabolism and subsequently intervening in the progression of NAFLD.

2.3.5 Tryptophan metabolites

Various metabolites derived from dietary tryptophan by the gut microbiota are emerging as essential markers for metabolic disorders [32]. Tryptophanase (TnaA), expressed in many gut bacteria, catalyzes the conversion of tryptophan into indole and its derivatives, such as indole-3-acetic acid (IAA) and indole-3-propionic acid (IPA) [33]. Bacteria like *Escherichia coli*, *Clostridium*, and *Bacteroides* can convert tryptophan into these metabolites [33]. Additionally, *Peptostreptococcus* species can convert tryptophan into indole acrylic acid (IAA) and IPA, while *Lactobacillus* and *Akkermansia* convert tryptophan into indole-3-acetaldehyde (IAld) and IAA [34, 35]. Furthermore, *Ruminococcus* is known to convert tryptophan into tryptamine, and *Bifidobacterium* has also been reported to produce Indole-3-lactic acid (ILA) [36, 37]. These metabolites also serve as potential targets for investigating the relationship between the gut microbiota and NAFLD.

3. Potential probiotics in preclinical studies

3.1 *Akkermansia muciniphila*

Akkermansia muciniphila (*A. muciniphila*) has demonstrated promising pharmacological effects in treating fatty liver disease by regulating metabolism, inflammation, and gut-liver interactions. It reduces body weight gain and fat mass and improves glucose tolerance and insulin sensitivity by modulating lipid metabolism in the liver and muscle, lowering fat synthesis, and alleviating endoplasmic reticulum (ER) stress [38]. Additionally, *A. muciniphila* decreases chronic inflammation by reducing plasma levels of LPS-binding protein (LBP) and leptin while increasing anti-inflammatory factors such as α-tocopherol and β-sitosterol in chow diet-fed mice [38]. Furthermore, *A. muciniphila* modulates gut microbiota composition, enhancing mitochondrial oxidation, bile acid metabolism, and L-aspartate metabolism [39], which collectively contribute to the mitigation of hepatic steatosis, inflammation, and oxidative stress. *A. muciniphila* reduces harmful gut species (e.g., *Alistipes*, *Lactobacilli*) and promotes beneficial ones (e.g., *Ruminiclostridium*, *Anaeroplasma*), thereby impacting bile acid metabolism through the FXR-FGF15 axis in high-fat-diet-related MAFLD [40]. This process reduces secondary bile acids that contribute to the alleviation of liver injury [40]. In NASH models, *A. muciniphila* also diminishes liver injury, steatosis, and inflammation by modulating immune responses, such as increasing CXCR6+ NKT cell activity, reducing macrophage infiltration, and inhibiting pro-inflammatory macrophage polarization [41]. Moreover, *A. muciniphila* prevented NASH by modulating TLR2-activated γδT17 cells and further macrophage polarization, providing protection against NASH progression [42]. Besides its role in immune modulation, *A. muciniphila* enhances antioxidant defense by restoring the GSH/GSSG balance and increasing superoxide dismutase (SOD) activity in mice APAP-induced liver injury models, thereby alleviating oxidative stress [43]. It also promotes SCFAs secretion, which improves gut barrier function and reduces inflammation. Furthermore, *A. muciniphila* influences gene expression related to the endocannabinoid system and peroxisome proliferator-activated receptors (PPARs) [44]. Notably, *A. muciniphila* supplementation reduces acetaminophen-induced liver injury by enhancing the gut barrier, restoring oxidative balance, and activating protective signaling pathways like PI3K/Akt [43]. Taken together, *A. muciniphila* is emerging as a promising therapeutic agent for metabolic liver diseases [45].

3.2 *Bacteroides* species

The genus *Bacteroides*, such as *Bacteroides ovatus*, *Bacteroides uniformis*, *Bacteroides thetaiotaomicron*, and *Bacteroides vulgatus*, was reported to exert pharmacological effects by influencing the gut-liver axis, altering host metabolism, and enhancing immune responses [46–51]. For instance, *B. ovatus* could alter the F/B ratio, promote beneficial bacteria like *Lachnospiraceae*, and reduce endotoxins in the HFHC diet model, which collectively improves liver function and alleviates liver inflammation. Additionally, *B. ovatus* enhances lipid metabolism by upregulating genes for fatty acid oxidation, such as *Ppara*, and downregulating lipogenesis genes *Srebf1* [46]. Similarly, *B. uniformis* improves glucose metabolism and reduces adiposity, particularly when combined with wheat bran extract. It enhances SCFAs production and boosts immune responses by increasing intraepithelial lymphocytes and type-3 innate lymphoid cells in obese mice [47]. *B. uniformis* also suppresses IL-22 signaling and liver inflammation, contributing to metabolic improvements [47, 52]. Furthermore, *B. thetaiotaomicron* enhances lipid metabolism by improving intestinal barrier function and reducing liver triglyceride accumulation in HFD-fed mice. It also restores key gut hormones GLP-1 and FGF15, which are vital for metabolic regulation and lipid metabolism [48, 49]. Lastly, *B. vulgatus* reduces hyperlipidemia and inflammation by modulating bile acid metabolism and SCFA production. It also alters the gut microbiota, thereby improving lipid metabolism and preventing fatty liver in HFHC or high-fat diet mice [50, 51]. Overall, *Bacteroides* species exhibited promising therapeutic potential for treating MASLD, although their effects may vary depending on strain-specific interactions with the host.

3.3 *Bacillus* species

Bacillus species have shown promise as probiotics for treating fatty liver disease by modulating lipid metabolism, regulating gut microbiota, and reducing inflammation. Several strains, such as *Bacillus toyonensis* SAU-20 and *Bacillus subtilis*, have been reported to alleviate hepatic steatosis and improve liver health [53–55]. *Bacillus toyonensis* SAU-20 alleviates hepatic steatosis and insulin resistance in type 2 diabetic mice by reducing hepatic fat, downregulating lipogenic genes and inflammatory cytokines, and alleviating oxidative stress, thereby improving liver function and lipid metabolism [53]. *Bacillus subtilis* improves liver health by restoring intestinal barrier function, reducing inflammatory cytokines, and altering gut microbiota composition, which could reduce liver fat accumulation and inflammation [54, 55]. Significantly, *Bacillus subtilis* HGCC-1 improves lipid metabolism by enhancing fatty acid β-oxidation and promoting lipid absorption and hepatic transport in golden pompanos [55]. In conclusion, *Bacillus* probiotics might also represent a promising approach for managing fatty liver disease through multiple mechanisms.

3.4 *Clostridium* species

Clostridium species, such as *Clostridium butyricum* and *Clostridium tyrobutyricum*, have emerged as promising candidates for treating MASLD [56–60]. These bacteria regulate lipid metabolism, modulate gut microbiota, and enhance liver health. One of the key therapeutic actions of *C. butyricum* CBM588 is its capacity to reduce lipid accumulation in the liver by downregulating triglyceride synthesis DGAT2 and simultaneously upregulating genes involved in cholesterol catabolism and bile acid excretion in HFD-induced fatty liver in rats. It also enhances the expression of PPARα/γ and LXRα, which are pivotal transcription factors in regulating lipid metabolism [57]. Similarly, *C. tyrobutyricum* has shown benefits in HFD-induced obese mice by reducing liver triglycerides, cholesterol, and non-esterified fatty acids. The mechanisms involve downregulating PPARγ and upregulating AMPK, PPARα, and lipid metabolism-related enzymes, such as ATGL and HSL, which could improve hepatic lipid profiles [60]. Additionally, modulation of the gut microbiota is crucial for metabolic health. *C. butyricum* reshapes the microbiota by increasing beneficial bacteria like *Lactobacillus* and producing butyrate, which enhances gut health, reduces inflammation, and provides additional liver protection. Moreover, both species have been shown to reduce oxidative stress and inflammatory markers, such as TNF-α and IL-6. In conclusion, *C. butyricum* and *C. tyrobutyricum* represent promising therapeutic strategies for MASLD.

3.5 *Parabacteroides* species

Parabacteroides species, such as *Parabacteroides distasonis*, *Parabacteroides goldsteinii*, and *Parabacteroides merdae*, have shown promise for the management of MASLD [61–65]. Their therapeutic effects primarily stem from the modulation of gut microbiota, bile acid metabolism, and inflammation. Notably, *P. distasonis* is particularly effective in altering bile acid profiles by increasing levels of lithocholic acid (LCA) and ursodeoxycholic acid (UDCA), which activate the FXR pathway in ob/ob and HFD-fed mice. This activation improves liver metabolism, maintains gut integrity, and reduces hepatic fat accumulation as well as hyperlipidemia [61]. Additionally, *P. distasonis* boosts bile salt hydrolase (BSH) activity, which reduces the toxic effects of taurochenodeoxycholic acid (TCDCA) on the liver in the TAA- and MCD diet-induced hepatic fibrosis mice model, thereby reducing hepatic stellate cell activation and alleviating liver fibrosis [62]. Similarly, *P. goldsteinii* contributes to alleviating obesity and insulin resistance, the common symptoms associated with fatty liver in HFD-fed mice. This bacterium also enhances adipose tissue thermogenesis, improves intestinal integrity, diminishes systemic inflammation, and modulates gut microbiota composition, which supports metabolic functions and reduces liver fat accumulation [64]. Moreover, *P. distasonis* used inulin to produce pentadecanoic acid, an essential odd-chain fatty acid, which restored gut barrier function and reduced serum lipopolysaccharide and liver pro-inflammatory cytokine expression in NASH models [63]. Besides, *P. merdae* can produce short-chain fatty acids that reduce inflammation and enhance lipid metabolism through the catabolism of branched-chain amino acids (BCAAs). Consequently, this process may help prevent hepatic steatosis by regulating the mTORC1 pathway on atherosclerosis in $ApoE^{-/-}$ mice [65]. In conclusion, *Parabacteroides* species, notably *P. distasonis*, *P. goldsteinii*, and *P. merdae*, exhibit therapeutic potential for MASLD by modulating bile acid metabolism, regulating gut microbiota, reducing inflammation, and promoting liver protection.

3.6 *Bifidobacterium* species

Bifidobacterium species exhibit considerable potential in treating MASLD through multiple mechanisms, including the modulation of lipid metabolism, reduction of inflammation, and enhancement of gut barrier integrity. Various strains of *Bifidobacterium* mediate these effects, with each strain targeting specific molecular pathways. For example, *B. animalis* subsp. *lactis* V9 reduces hepatic triglycerides and free fatty acids while simultaneously increasing glycogen levels, suggesting improved lipid storage in rats with NAFLD [66]. Similarly, *B. bifidum* enhances propionic and butyric acid production, which improves lipid metabolism and intestinal permeability, ultimately reducing liver fat accumulation and inflammation in HFD-fed mice [67]. Moreover, *B. longum* BL19 alleviates NAFLD by decreasing liver inflammation and modulating serum metabolites such as butyric acid, which reduces oxidative stress in mice with NAFLD [68]. *Bifidobacterium* strains also exhibit potent anti-inflammatory effects, further supporting NAFLD management. As for the inflammation regulation, *B. animalis* subsp. *lactis* V9 suppresses inflammatory cytokines by inhibiting pathways such as TLR4, TLR9, NLRP3, and ASC [66]. *B. adolescentis* preserves gut barrier function, reducing LPS levels and preventing liver inflammation through the TLR4/NF-kB signaling pathway in mice on a CDHFD [69]. Additionally, *B. lactis* SF exerts anti-inflammatory pathways through the TLR4/NF-κB and PI3K-Akt/AMPK pathways in the NAFLD mouse model, while

B. pseudolongum significantly suppressed NAFLD-HCC progression by secreting acetate, which bound to hepatic GPR43 (G-coupled protein receptor 43) via the gut-liver axis and suppressed the oncogenic IL-6/JAK1/STAT3 signaling pathway [70, 71]. These strains also regulate the gut-liver axis by maintaining the integrity of the gut barrier. For example, *B. lactis* SF restores intestinal flora and prevents LPS translocation into the liver, while *B. animalis* subsp. *lactis* B420 significantly alleviated S100-induced experimental autoimmune hepatitis (EAH) by enhancing intestinal barrier function and reducing endotoxins, improving liver function [70, 72]. Additionally, *B. longum* and *B. bifidum* enhance bile acid signaling, which improves oxidative phosphorylation and regulates metabolism, thereby protecting against obesity and fatty liver disease. *B. lactis* MG741 influences lipid metabolism by modulating key enzymes such as ACC, FAS, and SREBP-1 in HFD-fed mice [72]. In conclusion, *Bifidobacterium* species also present a multifaceted therapeutic potential for NAFLD through mechanisms that include modulation of lipid metabolism, enhancement of gut barrier function, regulation of metabolites, and reduction of inflammation.

3.7 Lactic acid bacteria

Lactic acid bacteria, including genera such as *Lactobacillus*, *Lactococcus*, *Limosilactobacillus*, and *Lacticaseibacillus,* showed significant therapeutic potential in managing MASLD by targeting liver metabolism, inflammation, oxidative stress, and gut microbiota composition. Firstly, these bacteria effectively regulate lipid metabolism. For instance, *Lactobacillus plantarum* NA136 activates the AMPK pathway, which subsequently phosphorylates ACC, thereby inhibiting de novo lipogenesis and promoting fatty acid oxidation on HFD and fructose (HFD/F)-induced NAFLD [73]. Similarly, *Lactobacillus acidophilus* YL01 regulates lipid metabolism through the AMPK/ACC pathway, reducing body weight and fat accumulation in high-fat mice. It also increases beneficial bacteria that produce SCFAs, thus enhancing gut-liver cross-talk [74]. Moreover, *Lactobacillus salivarius* SNK-6 regulates lipid metabolism through the miR-130a-5p/MBOAT2 signaling pathway, resulting in decreased liver triglycerides and improved liver enzyme levels, while also modulating bile acids such as cholic and ursodeoxycholic acids in a NAFLD model [75]. Inflammation is another central factor in MASLD progression, and several lactic acid bacteria exhibit anti-inflammatory properties. *Lactobacillus helveticus* R0052 inhibits hepatic pro-inflammatory cytokines such as TNF-α and NF-κB in D-galactosamine-treated rats, while *Lactobacillus plantarum* C88 protects against aflatoxin B1-induced liver injury in mice via inhibition of NF-κB-mediated inflammatory responses and excessive apoptosis [76, 77]. Additionally, lactic acid bacteria contribute to the attenuation of oxidative stress. *Lactobacillus reuteri* DSM 17938 reduces oxidative stress and improves liver function in drug-induced liver injury (DILI) by activating the AMPK pathway and enhancing mitochondrial biogenesis and energy supply [78]. Similarly, *Lactobacillus rhamnosus* JL1 upregulates fatty acid oxidation and downregulates lipogenesis-related genes, reducing oxidative stress in high-fat diet mice [79]. Lactic acid bacteria also positively influence gut microbiota composition. For example, *Lactobacillus plantarum* NA136 modulates the microbiome by increasing beneficial bacteria *Bifidobacterium* and *Lactobacillus* and decreasing pathogenic species, thereby improving gut health and reducing systemic inflammation [73]. *Lactobacillus reuteri* CCFM8631 also increases beneficial bacteria and reduces harmful ones, improving lipid metabolism and alleviating hypercholesterolemia in mice fed a Paigen atherogenic diet [80]. In terms of metabolic health, lactic acid bacteria also improve insulin sensitivity. *Lactobacillus plantarum* NA136 alleviates insulin resistance and reduces hepatic lipid content in NAFLD mice [73]. *Lactobacillus gasseri* BNR17 reduces body weight, fat accumulation, and liver inflammation while enhancing glucose tolerance in obese mice [81]. Furthermore, lactic acid bacteria show promise in preventing liver fibrosis. *Lactobacillus reuteri* GMNL-263 reduces liver and heart fibrosis in high-fat diet-induced hamsters by reducing lipid metabolic stress and suppressing TGF-β expression [82]. *Lactobacillus salivarius* SNK-6 also alleviates liver fibrosis in an NAFLD chicken model through metabolic modulation and increased SCFA production [75]. Additionally, *L. acidophilus* exerts a protective function against NAFLD-HCC by producing valeric acid, which enhances intestinal barrier integrity and inhibits the Rho-GTPase pathway via the GPR41/43 receptor on hepatocytes [83]. *Lactobacillus reuteri* CCFM8631 reduces plasma cholesterol and liver triglycerides while simultaneously increasing liver superoxide dismutase (SOD) levels, which aids in alleviating oxidative stress [80]. *Lactobacillus fermentum* TY-S11 prevents the loss of gut microbiota diversity and promotes SCFA production in high-cholesterol diet-fed apolipoprotein E-deficient mice [84, 85]. In conclusion, lactic acid bacteria offer promising therapeutic potential in managing MASLD through multiple mechanisms, including the regulation of lipid metabolism, inflammation, gut microbiota, insulin sensitivity, liver fibrosis, and cholesterol content.

3.8 Others

Several other bacteria have shown promise in treating MASLD through various mechanisms. For example, *Faecalibacterium prausnitzii* has demonstrated potential in treating NASH by regulating gut microbiota, improving gut barrier function, and reducing hepatic lipid accumulation, fibrosis, and liver damage [86]. Furthermore, *F. prausnitzii* promotes fatty acid oxidation, enhances adiponectin signaling, and improves insulin sensitivity in adipose tissues, boosting mitochondrial respiration and altering gut microbiota composition, thereby benefiting liver and adipose tissue health [87]. Research also indicates that *F. prausnitzii* restores serum lipid profiles and reduces oxidative stress by modulating gut microbiota, which enhances tryptophan and branched-chain amino acid metabolism [88]. Furthermore, *Desulfovibrio vulgaris*, enriched by APS from *Astragalus mongholicus*, produces acetic acid that inhibits FASN and CD36, leading to alleviated hepatic steatosis in HFD-fed mice [89].

In contrast, *Roseburia* species, particularly *R. intestinalis*, could enhance gut barrier function, reduce intestinal permeability, and alleviate hepatic inflammation and steatosis [90]. *Enterococcus faecalis* EF-2001 reduces hepatic lipid deposition and alleviates liver function in diet-induced obese mice by activating the AMPK pathway and downregulating lipid accumulation-related proteins [91, 92]. *Enterococcus faecium* GEFA01 shows cholesterol-lowering effects by modulating gut microbiota and enhancing cholesterol metabolism in the liver [93]. Lastly, *Extibacter muris* does not directly affect cholesterol but could produce DCA, which influences liver proteomics and benefits lipid homeostasis [94]. Overall, these strains highlight the therapeutic potential of gut microbiota modulation in improving lipid metabolism and liver health.

4. Application of probiotics for MASLD in clinical trials

The current clinical application of probiotic therapy for MASLD (NAFLD) could be summarized as single-strain probiotics, multistrain probiotics, and synbiotics (**Table 1**). These representative strains have demonstrated varying degrees of efficacy

Component	Outcomes	Objects	Ref.
Single-strain probiotics			
Lactobacillus rhamnosus GG	Serum LT, PG-PS IgA↓	Obese children	[95]
Bacillus coagulans	Serum ALT, GGT, TNF-α, NF-κB activity; hepatic steatosis↓	NAFLD patients	[96]
Bacillus coagulans TCI711	Fatty liver grade↓	NAFLD patients	[97]
Bacillus clausii	Blood glucose↓, plasma HDL↑	T2DM patients	[98]
Bifidobacterium animalis subsp. *lactis* CECT 8145	CECT 8145: WC, Conicity index, BMI↓, Heat killed CECT 8145: VFA↓, DBP, HOMA index↓	Abdominally obese subjects	[99]
Bifidobacterium breve BBr60	Weight, BMI, FBG↓	Overweight or obese adults	[100]
Bifidobacterium lactis IDCC 4301	Total fat mass, LAR, BMI, leg fat↓	Volunteers with obesity	[101]
Lactobacillus sakei CJLS03	Body fat mass, WC↓	Adults with obesity	[102]
Lactobacillus sakei OK67	VFA↓	Overweight individuals	[103]
Lactobacillus rhamnosus CGMCC1.3724	Weight loss, fat mass, leptin in women↓,	Obese men and women	[104]
Lactobacillus reuteri V3401	Plasma IL-6, sVCAM-1↓	Adults with MetS	[105]
Lactobacillus plantarum LMT1-48	Body weight, VFA↓; serum insulin, IR, leptin↓	Volunteers with obesity	[106]
Lactobacillus gasseri SBT2055	VFA, subcutaneous fat areas, body weight, BMI, waist, hip↓; serum HMW adiponectin↑	Adults with obese tendencies	[107]
Lactobacillus gasseri BNR17	WC, VAT↓	Obese adults	[108]
Lactobacillus casei Shirota	Body weight↓; Serum HDL-cholesterol↑	Obese children	[109]
Lacticaseibacillus rhamnosus HA-114	Plasma insulin, HOMA-IR, LDL-cholesterol, TC↓	Adults with overweight	[110]
Multistrain probiotics			
VSL#3: *Streptococcus thermophilus*, *Bifidobacteria breve*, *Bifidobacteria infantis*, *Bifidobacteria longum*, *Lactobacillus acidophilus*, *Lactobacillus plantarum*, *Lactobacillus paracasei*, *Lactobacillus delbrueckii* subsp. *bulgaricus*	BMI↓; Serum GLP-1, aGLP1↑	Children with NAFLD	[111]
Lepicol probiotic: *Lactobacillus plantarum*, *Lactobacillus delbrueckii*, *Lactobacillus acidophilus*, *Lactobacillus rhamnosus*, *Bifidobacterium bifidum*	IHTG, Serum AST↓	NASH patients	[112]

Component	Outcomes	Objects	Ref.
"Symbiter": *Lactobacillus* spp., *Lactococcus* spp., *Bifidobacterium* spp., *Propionibacterium* spp.	FLI; Serum AST, GGT, TNF-*α*, IL-6↓	NAFLD Patients	[113]
MCP® BCMC® strains: *Lactobacillus* spp., *Bifidobacterium* spp.	BMI, Fibrosis score, Fasting glucose↓	NAFLD Patients	[114]
Lactobacillus acidophilus, *Bifidobacterium breve*, *Bifidobacterium longum*, *Lactobacillus delbrueckii* susp. *Bulgaricus*, *Lactobacillus helveticus*, *Lactobacillus plantarum*, *Lactobacillus rhamnosus*, *Lactobacillus casei*, *Lactococcus lactis* susp. *Lactis*, *Streptococcus thermophiles*	NAFLD fibrosis score, VAI; Serum AST, TG↓	Obese patients	[115]
Lactobacillus acidophilus, *Bifidobacterium lactis*	APRI score↓	NASH patients	[116]
Lactobacillus casei, *Lactobacillus acidophilus*, *Lactobacillus rhamnosus*, *Lactobacillus bulgaricus*, *Bifidobacterium breve*, *Bifidobacterium longum*, *Streptococcus thermophilus*	FBS; Serum insulin, IR, IL-6↓	NAFLD patients	[117]
Lactobacillus acidophilus CBT LA1, *Lactobacillus rhamnosus* CBT LR5, *Lactobacillus paracasei* CBT LPC5, *Pediococcus pentosaceus* CBT SL4, *Bifidobacterium lactis* CBT BL3, *Bifidobacterium breve* CBT BR3	VFA, IHF fraction, Serum TG↓	Obese NAFLD patients	[118]
Bifidobacterium breve BR03 and B632 strains	Weight↓ insulin sensitivity at fasting↑	Obesity and insulin resistance on diet	[119]
Lacticaseibacillus paracasei BEPC22, *Lactiplantibacillus plantarum* BELP53	Body fat index, body weight↓	Overweight participants	[120]
Lactobacillus curvatus HY7601, *Lactobacillus plantarum* KY1032	Body weight, visceral fat mass, WC↓ adiponectin↑	Overweight individuals	[121]
LactoLevureR: *Lactobacillus acidophilus*, *Lactobacillus plantarum*, *Bifidobacterium lactis*, *Saccharomyces boulardii*	Blood HbA1c, FBG, TC↓	Adults with T2D	[122]
Symbiter: *Lactobacillus*, *Lactococcus*, *Bifidobacterium*, *Propionibacterium*, *Acetobacter*	HOMA-IR↓; Serum HbA1c, TNF-α, IL-1β↓	Patients with T2D	[123]
Synbiotics			
Lactobacillus casei, *Lactobacillus rhamnosus*, *Streptococcus thermophilus*, *Bifidobacterium breve*, *Lactobacillus acidophilus*, *Bifidobacterium longum*, *Lactobacillus bulgaricus* FOS	Hepatic steatosis↓; serum FBS, TAG, AST↓	NAFLD patients	[124]

Component	Outcomes	Objects	Ref.
Lactobacillus casei, *Lactobacillus rhamnosus*, *Lactobacillus bulgaricus*, *Bifidobacterium longum*, *Streptococcus thermophilus* FOS	Liver stiffness, BMI, Serum ALT, Chol↓	NASH patients	[125]
Lactobacillus casei, *Lactobacillus rhamnosus*, *Lactobacillus acidophilus*, *Lactobacillus bulgaricus*, *Bifidobacterium breve*, *Bifidobacterium longum*, *Streptococcus thermophilus* FOS	Serum FBS, Chol, LDL, AST↓	NAFLD patients	[126]
Lactobacillus casei, *Lactobacillus rhamnosus*, *Streptococcus thermophilus*, *Bifidobacterium breve*, *Lactobacillus acidophilus*, *Bifidobacterium longum*, *Lactobacillus bulgaricus* FOS	Serum ALT, AST, GGT, hs-CRP, TNF-α, NF-κB p65↓; fibrosis score↓	NAFLD patients	[127]
Bifidobacterium animalis subsp. *lactis* GCL2505 Inulin	REE score↑	Overweight or mildly obese adults	[128]
Lactobacillus acidophilus NCFM, *Bifidobacterium lactis* HN019 Polydextrose	Fasting glucose, insulin, HOMA-IR, TG↓ HDL-cholesterol↑	Overweight or obese individuals	[129]
Saccharomyces boulardii SOD	Body weight, BMI, fat mass, HOMA Index↓ Serum insulin, UA↓	Obese adults	[130]
Lactobacillus reuteri Guar gum and inulin	Steatosis, weight, BMI, WC↓	NASH patients	[131]

Table 1.
Clinical application of probiotics.

in alleviating liver function, inflammation, insulin resistance, and hepatic steatosis in patients with fatty liver. The clinical effects of single-strain probiotics, such as *Lactobacillus rhamnosus* GG, *Lactobacillus casei*, and *Bacillus* species, include reductions in liver function indicators such as Alanine Aminotransferase (ALT), Aspartate Aminotransferase (AST), γ-Glutamyl Transferase (GGT), Alkaline Phosphatase (ALP), and bilirubin, as well as alterations in gut microbiota profiles, including *Bacteroidota*, *Actinomycetota*, and *Bacillota* [95–97, 132]. Additionally, these strains have been found to decrease serum levels of LPS and TNF-α in NAFLD patients. Multistrain probiotics, such as those that combine multistrain including *Lactobacillus* spp. and *Bifidobacterium* spp. et al., are associated with improvements in hepatic steatosis, liver stiffness, metabolic liver function, and inflammatory indicators [111, 133]. Regarding synbiotics, the predominant prebiotic is fructo-oligosaccharide, which theoretically could positively impact host health by promoting the growth of beneficial bacteria, improving intestinal health, enhancing mineral absorption, and regulating immune function. The beneficial effects of these synbiotics are also primarily observed in decreased liver function, improved systemic metabolic parameters, and a balanced microbial community.

Although these preliminary clinical applications of probiotics on MASLD demonstrate varying degrees of therapeutic effects, they are far from satisfying, with inconsistent results across studies and, in some cases, no significant improvements observed. For example, in a study on *Bifidobacterium breve* Bif195, no significant differences in intestinal permeability or gastrointestinal symptoms were observed between the Bif195 and placebo groups, suggesting that its protective effects may not be primarily related to small bowel permeability in exercise-induced challenges [134]. Similarly, in a randomized placebo-controlled trial of the VSL#3® probiotic (a mixture of eight probiotic strains), no significant improvement was found in cardiovascular biomarkers or liver injury in patients with NAFLD, although a correlation between endothelial dysfunction and inflammation was noted [133]. However, a randomized clinical trial on obese children with NASH showed that VSL#3 significantly improved the severity of NAFLD, which was evidenced by a marked reduction in fatty liver severity and an increase in GLP-1 levels, which may contribute to the beneficial effects observed in this population [111]. Therefore, large-scale, randomized, controlled trials are urgently needed to identify the most effective probiotics or symbiotics.

5. Conclusions

From the current evidence, regulating the microbiome and the gut-liver axis appears to be a promising strategy for treating MASLD. Certain potential probiotics reported in animal studies, such as *A. muciniphila*, *Lactobacillus* strains, and *Bifidobacterium* species, have demonstrated particularly beneficial effects, while other less-studied probiotics like *Roseburia* and *Faecalibacterium* also show potential. The beneficial effects of these potential probiotics are likely due to alterations in gut microbiota, improved intestinal barrier integrity, and changes in microbial metabolites, which contribute to the enhancements in liver function, reductions in fat accumulation, decreased systemic inflammation, and even fibrosis. However, the effects observed in animal models do not directly translate to satisfied clinical outcomes, and challenges such as small sample sizes, strain variability, and a lack of long-term data persist. Consequently, there is an urgent need for large-scale, randomized, controlled trials to identify the solid, effective probiotic strains, combinations, or synbiotics for MASLD. Furthermore, additional research is essential to deepen our understanding of the underlying mechanisms through which probiotics affect the gut-liver axis. It is also necessary to discover new probiotic strains with specific therapeutic properties based on microbiome sequencing techniques, develop personalized and strain-specific probiotic therapy schemes, and explore combination therapy strategies in the future.

Acknowledgements

This work was supported by the CAMS Innovation Fund for Medical Sciences [2021-I2M-1-028 and 2021-I2M-1-055] and the National Natural Science Foundation of China (No. 82173722).

Conflict of interest

The authors declare no conflict of interest.

Abbreviations

aGLP-1	activated GLP-1
ALT	alanine transaminase
APRI	AST to platelet ratio index
AST	aspartate aminotransferase
BAs	bile acids
BMI	body mass index
BMISDS	body mass index standard deviation score
Chol	cholesterol
DBP	diastolic blood pressure
F/B ratio	firmicutes/bacteroidetes ratio
FBG	fasting blood glucose
FBS	fasting blood sugar
FOS	fructooligosaccharides
GGT	γ glutamine transaminase
GLP-1	glucagon-like peptide 1
HDL	high density lipoprotein cholesterol
HMW	high molecular weight
HOMA-IR	homeostasis model assessment estimated insulin resistance
hs-CRP	high-sensitivity C-reactive protein
IHTG	intrahepatic triglyceride content
IL-6	interleukin-6
IR	Insulin resistance
LAR	log leptin/adiponectin ratio
LDL	low density lipoprotein
LPS	lipopolysaccharide
MASH	metabolic dysfunction-associated steatohepatitis;
MASLD	metabolic dysfunction-associated steatotic liver disease
MetS	metabolic syndrome
NAFLD	nonalcoholic fatty liver disease
NASH	nonalcoholic steatohepatitis
NF-κB	nuclear factor-kB
PG-PS IgA	peptidoglycan-polysaccharide immunoglobulin A.
REE	resting energy expenditure
SCFAs	short-chain fatty acids
SOD	superoxide dismutase
sVCAM-1	soluble vascular cell adhesion molecule 1
TC	total cholesterol
TG	triglycerides
UA	uric acid
VAI	visceral adiposity index
VFA	abdominal visceral fat area
WC	waist circumference
WHtR	waist-height ratio

Author details

Xiayun Li[1], Liyan Yu[1], Zonggen Peng[1], Tingxia Lv[2] and Hu Li[1]*

1 Institute of Medicinal Biotechnology, Chinese Academy of Medical Sciences and Peking Union Medical College, Beijing, China

2 Department of Infectious Diseases, Beijing Friendship Hospital, Capital Medical University, China

*Address all correspondence to: lihu0112@163.com

References

[1] Ali SMJ et al. Metabolic dysfunction-associated steatotic liver disease of article. Annals of Internal Medicine. 2025;**178**(1):Itc1-itc16. DOI: 10.7326/annals-24-02933

[2] Jiang J et al. Upregulation of hepatic glutathione S-transferase alpha 1 ameliorates metabolic dysfunction-associated steatosis by degrading fatty acid binding protein 1 of article. International Journal of Molecular Sciences. 2024;**25**:10. DOI: 10.3390/ijms25105086

[3] Hsu CL et al. The gut-liver axis and gut microbiota in health and liver disease of article. Nature Reviews. Microbiology. 2023;**21**(11):719-733. DOI: 10.1038/s41579-023-00904-3

[4] Lang S et al. Microbiota and fatty liver disease-the known, the unknown, and the future of article. Cell Host & Microbe. 2020;**28**(2):233-244. DOI: 10.1016/j.chom.2020.07.007

[5] Marshall JC. The gut as a potential trigger of exercise-induced inflammatory responses of article. Canadian Journal of Physiology and Pharmacology. 1998;**76**(5):479-484. DOI: 10.1139/cjpp-76-5-479

[6] Stanislawski MA et al. Gut microbiota in adolescents and the association with fatty liver: The EPOCH study of article. Pediatric Research. 2018;**84**(2):219-227. DOI: 10.1038/pr.2018.32

[7] Gupta H et al. Gut microbiome in non-alcoholic fatty liver disease: From mechanisms to therapeutic role of article. Biomedicine. 2022;**10**:3. DOI: 10.3390/biomedicines10030550

[8] Nakano H et al. Bilberry anthocyanins ameliorate NAFLD by improving dyslipidemia and gut microbiome dysbiosis of article. Nutrients. 2020;**12**:11. DOI: 10.3390/nu12113252

[9] Del Chierico F et al. Gut microbiota profiling of pediatric nonalcoholic fatty liver disease and obese patients unveiled by an integrated meta-omics-based approach of article. Hepatology. 2017;**65**(2):451-464. DOI: 10.1002/hep.28572

[10] Moreira GV et al. Liraglutide modulates gut microbiota and reduces NAFLD in obese mice of article. The Journal of Nutritional Biochemistry. 2018;**62**:143-154. DOI: 10.1016/j.jnutbio.2018.07.009

[11] Spadoni I et al. Organ-specific protection mediated by cooperation between vascular and epithelial barriers of article. Nature Reviews. Immunology. 2017;**17**(12):761-773. DOI: 10.1038/nri.2017.100

[12] Compare D et al. Gut–liver axis: The impact of gut microbiota on non alcoholic fatty liver disease of article. Nutrition, Metabolism, and Cardiovascular Diseases. 2012;**22**(6):471-476. DOI: 10.1016/j.numecd.2012.02.007

[13] Soares JB et al. The role of lipopolysaccharide/toll-like receptor 4 signaling in chronic liver diseases of article. Hepatology International. 2010;**4**(4):659-672. DOI: 10.1007/s12072-010-9219-x

[14] Ferro D et al. New insights into the pathogenesis of non-alcoholic fatty liver disease: Gut-derived lipopolysaccharides and oxidative stress of article. Nutrients. 2020;**12**:9. DOI: 10.3390/nu12092762

[15] Bischoff SC et al. Intestinal permeability–A new target for disease

prevention and therapy of article. BMC Gastroenterology. 2014;**14**:189. DOI: 10.1186/s12876-014-0189-7

[16] Postler TS et al. Understanding the Holobiont: How microbial metabolites affect human health and shape the immune system of article. Cell Metabolism. 2017;**26**(1):110-130. DOI: 10.1016/j.cmet.2017.05.008

[17] Zhang S et al. Dietary fiber-derived short-chain fatty acids: A potential therapeutic target to alleviate obesity-related nonalcoholic fatty liver disease of article. Obesity Reviews. 2021;**22**(11):e13316. DOI: 10.1111/obr.13316

[18] Derrien M et al. *Akkermansia muciniphila* gen. Nov., sp. Nov., a human intestinal mucin-degrading bacterium of article. International Journal of Systematic and Evolutionary Microbiology. 2004;**54**(Pt. 5):1469-1476. DOI: 10.1099/ijs.0.02873-0

[19] Chambers ES et al. Role of gut microbiota-generated short-chain fatty acids in metabolic and cardiovascular health of article. Current Nutrition Reports. 2018;**7**(4):198-206. DOI: 10.1007/s13668-018-0248-8

[20] Nguyen HT et al. Lactobacillus sakei MJM60958 as a potential probiotic alleviated non-alcoholic fatty liver disease in mice fed a high-fat diet by modulating lipid metabolism, inflammation, and gut microbiota of article. International Journal of Molecular Sciences. 2022;**23**:21. DOI: 10.3390/ijms232113436

[21] Lu Y et al. Short chain fatty acids prevent high-fat-diet-induced obesity in mice by regulating G protein-coupled receptors and gut microbiota of article. Scientific Reports. 2016;**6**:37589. DOI: 10.1038/srep37589

[22] Molinero N et al. Intestinal bacteria interplay with bile and cholesterol metabolism: Implications on host physiology of article. Frontiers in Physiology. 2019;**10**:185. DOI: 10.3389/fphys.2019.00185

[23] Xue R et al. Bile acid receptors and the gut-liver axis in nonalcoholic fatty liver disease of article. Cells. 2021;**10**:11. DOI: 10.3390/cells10112806

[24] Wahlstrom A et al. Intestinal crosstalk between bile acids and microbiota and its impact on host metabolism of article. Cell Metabolism. 2016;**24**(1):41-50. DOI: 10.1016/j.cmet.2016.05.005

[25] Miao RR et al. Intestinal aberrant sphingolipid metabolism shaped-gut microbiome and bile acids metabolome in the development of hepatic steatosis of article. The FASEB Journal. 2022;**36**(8):e22398. DOI: 10.1096/fj.202200148RR

[26] Clifford BL et al. FXR activation protects against NAFLD via bile-acid-dependent reductions in lipid absorption of article. Cell Metabolism. 2021;**33**(8):1671-84 e4. DOI: 10.1016/j.cmet.2021.06.012

[27] Dai X et al. Microbial metabolites: Critical regulators in NAFLD of article. Frontiers in Microbiology. 2020;**11**:567654. DOI: 10.3389/fmicb.2020.567654

[28] Ruuskanen MO et al. Links between gut microbiome composition and fatty liver disease in a large population sample of article. Gut Microbes. 2021;**13**(1):1-22. DOI: 10.1080/19490976.2021.1888673

[29] Mbaye B et al. Endogenous ethanol and triglyceride production by gut *Pichia kudriavzevii*, *Candida albicans* and *Candida glabrata* yeasts in non-alcoholic

Steatohepatitis of article. Cells. 2022;**11**:21. DOI: 10.3390/cells11213390

[30] Aragones G et al. Gut microbiota-derived mediators as potential markers in nonalcoholic fatty liver disease of article. BioMed Research International. 2019;**2019**:8507583. DOI: 10.1155/2019/8507583

[31] Lin D et al. Gut microbiota and bile acids partially mediate the improvement of fibroblast growth factor 21 on methionine-choline-deficient diet-induced non-alcoholic fatty liver disease mice of article. Free Radical Biology & Medicine. 2023;**195**:199-218. DOI: 10.1016/j.freeradbiomed.2022.12.087

[32] Sehgal R et al. Indolepropionic acid, a gut bacteria-produced tryptophan metabolite and the risk of type 2 diabetes and non-alcoholic fatty liver disease of article. Nutrients. 2022;**14**:21. DOI: 10.3390/nu14214695

[33] Russell WR et al. Major phenylpropanoid-derived metabolites in the human gut can arise from microbial fermentation of protein of article. Molecular Nutrition & Food Research. 2013;**57**(3):523-535. DOI: 10.1002/mnfr.201200594

[34] Wlodarska M et al. Indoleacrylic acid produced by commensal *Peptostreptococcus* species suppresses inflammation of article. Cell Host & Microbe. 2017;**22**(1):25-37 e6. DOI: 10.1016/j.chom.2017.06.007

[35] Yang W et al. Fuzhuan brick tea polysaccharide improved ulcerative colitis in association with gut microbiota-derived tryptophan metabolism of article. Journal of Agricultural and Food Chemistry. 2021;**69**(30):8448-8459. DOI: 10.1021/acs.jafc.1c02774

[36] Saraf MK et al. Formula diet driven microbiota shifts tryptophan metabolism from serotonin to tryptamine in neonatal porcine colon of article. Microbiome. 2017;**5**(1):77. DOI: 10.1186/s40168-017-0297-z

[37] Hashikura N et al. Synbiotics of bifidobacterium breve MCC1274 and lactulose enhances production of tryptophan metabolites in fermented human fecal communities of article. Food Research International. 2023;**163**:112308. DOI: 10.1016/j.foodres.2022.112308

[38] Zhao S et al. *Akkermansia muciniphila* improves metabolic profiles by reducing inflammation in chow diet-fed mice of article. Journal of Molecular Endocrinology. 2017;**58**(1):1-14. DOI: 10.1530/JME-16-0054

[39] Rao Y et al. Gut *Akkermansia muciniphila* ameliorates metabolic dysfunction-associated fatty liver disease by regulating the metabolism of L-aspartate via gut-liver axis of article. Gut Microbes. 2021;**13**(1):1-19. DOI: 10.1080/19490976.2021.1927633

[40] Wu W et al. *Akkermansia muciniphila* alleviates high-fat-diet-related metabolic-associated fatty liver disease by modulating gut microbiota and bile acids of article. Microbial Biotechnology. 2023;**16**(10):1924-1939. DOI: 10.1111/1751-7915.14293

[41] Li T et al. *Akkermansia muciniphila* suppressing nonalcoholic steatohepatitis associated tumorigenesis through CXCR6(+) natural killer T cells of article. Frontiers in Immunology. 2022;**13**:1047570. DOI: 10.3389/fimmu.2022.1047570

[42] Han Y et al. *Akkermansia muciniphila* inhibits nonalcoholic steatohepatitis by orchestrating

TLR2-activated gammadeltaT17 cell and macrophage polarization of article. Gut Microbes. 2023;**15**(1):2221485. DOI: 10.1080/19490976.2023.2221485

[43] Xia J et al. *Akkermansia muciniphila* ameliorates acetaminophen-induced liver injury by regulating gut microbial composition and metabolism of article. Microbiol Spectr. 2022;**10**(1):e0159621. DOI: 10.1128/spectrum.01596-21

[44] Ghaderi F et al. Effects of active, inactive, and derivatives of *Akkermansia muciniphila* on the expression of the endocannabinoid system and PPARs genes of article. Scientific Reports. 2022;**12**(1):10031. DOI: 10.1038/s41598-022-13840-8

[45] Niu H et al. *Akkermansia muciniphila*: A potential candidate for ameliorating metabolic diseases of article. Frontiers in Immunology. 2024;**15**:1370658. DOI: 10.3389/fimmu.2024.1370658

[46] Sun C et al. Bacteroides ovatus alleviates high-fat and high-cholesterol-induced nonalcoholic fatty liver disease via gut-liver axis of article. Biomedicine & Pharmacotherapy. 2024;**178**:117156. DOI: 10.1016/j.biopha.2024.117156

[47] Lopez-Almela I et al. *Bacteroides uniformis* combined with fiber amplifies metabolic and immune benefits in obese mice of article. Gut Microbes. 2021;**13**(1):1-20. DOI: 10.1080/19490976.2020.1865706

[48] Cho SH et al. The human symbiont *Bacteroides thetaiotaomicron* promotes diet-induced obesity by regulating host lipid metabolism of article. Journal of Microbiology. 2022;**60**(1):118-127. DOI: 10.1007/s12275-022-1614-1

[49] Li H et al. *Bacteroides thetaiotaomicron* ameliorates mouse hepatic steatosis through regulating gut microbial composition, gut-liver folate and unsaturated fatty acids metabolism of article. Gut Microbes. 2024;**16**(1):2304159. DOI: 10.1080/19490976.2024.2304159

[50] Xu M et al. *Bacteroides vulgatus* ameliorates lipid metabolic disorders and modulates gut microbial composition in Hyperlipidemic rats of article. Microbiology Spectrum. 2023;**11**(1):e0251722. DOI: 10.1128/spectrum.02517-22

[51] Jin S et al. *Phocaeicola vulgatus* alleviates diet-induced metabolic dysfunction-associated steatotic liver disease progression by downregulating histone acetylation level via 3-HPAA of article. Gut Microbes. 2024;**16**(1):2309683. DOI: 10.1080/19490976.2024.2309683

[52] Morita H et al. *Bacteroides uniformis* and its preferred substrate, α-cyclodextrin, enhance endurance exercise performance in mice and human males of article. Science Advances. 2023;**9**:4 eadd2120. DOI: 10.1126/sciadv.add2120

[53] Ren Z et al. Oral administration of *Bacillus toyonensis* strain SAU-20 improves insulin resistance and ameliorates hepatic steatosis in type 2 diabetic mice of article. Frontiers in Immunology. 2022;**13**:837237. DOI: 10.3389/fimmu.2022.837237

[54] Liu D et al. Binary *Bacillus subtilis* protects the intestinal mucosa barrier and alleviates nonalcoholic steatohepatitis of article. Animal Models and Experimental Medicine. 2024;**7**(3):362-366. DOI: 10.1002/ame2.12337

[55] Li M et al. Bacillus subtilis HGCC-1 improves growth performance and liver health via regulating gut microbiota in golden pompano of article. Animal

Microbiome. 2025;7(1):7. DOI: 10.1186/s42523-024-00372-x

[56] Zhang X et al. Supplemental *Clostridium butyricum* modulates lipid metabolism by reshaping the gut microbiota composition and bile acid profile in IUGR suckling piglets of article. Journal of Animal Science and Biotechnology. 2023;**14**(1):36. DOI: 10.1186/s40104-023-00828-1

[57] Seo M et al. *Clostridium butyricum* MIYAIRI 588 improves high-fat diet-induced non-alcoholic fatty liver disease in rats of article. Digestive Diseases and Sciences. 2013;**58**(12):3534-3544. DOI: 10.1007/s10620-013-2879-3

[58] Choi Y et al. Effect of *Clostridium butyricum* on high-fat diet-induced intestinal inflammation and production of short-chain fatty acids of article. Digestive Diseases and Sciences. 2023;**68**(6):2427-2440. DOI: 10.1007/s10620-023-07835-2

[59] Liu Y et al. A discovery of relevant hepatoprotective effects and underlying mechanisms of dietary clostridium butyricum against corticosterone-induced liver injury in Pekin ducks of article. Microorganisms. 2019;7:9. DOI: 10.3390/microorganisms7090358

[60] Luo Y et al. Effects of clostridium tyrobutyricum on lipid metabolism, intestinal barrier function, and gut microbiota in obese mice induced by high-fat diet of article. Nutrients. 2024;**16**:4. DOI: 10.3390/nu16040493

[61] Wang K et al. Parabacteroides distasonis alleviates obesity and metabolic dysfunctions via production of succinate and secondary bile acids of article. Cell Reports. 2019;**26**(1):222-35.e5. DOI: 10.1016/j.celrep.2018.12.028

[62] Zhao Q et al. Parabacteroides distasonis ameliorates hepatic fibrosis potentially via modulating intestinal bile acid metabolism and hepatocyte pyroptosis in male mice of article. Nature Communications. 2023;**14**(1):1829. DOI: 10.1038/s41467-023-37459-z

[63] Wei W et al. Parabacteroides distasonis uses dietary inulin to suppress NASH via its metabolite pentadecanoic acid of article. Nature Microbiology. 2023;**8**(8):1534-1548. DOI: 10.1038/s41564-023-01418-7

[64] Wu TR et al. Gut commensal *Parabacteroides goldsteinii* plays a predominant role in the anti-obesity effects of polysaccharides isolated from Hirsutella sinensis of article. Gut. 2019;**68**(2):248-262. DOI: 10.1136/gutjnl-2017-315458

[65] Qiao S et al. Gut *Parabacteroides merdae* protects against cardiovascular damage by enhancing branched-chain amino acid catabolism of article. Nature Metabolism. 2022;**4**(10):1271-1286. DOI: 10.1038/s42255-022-00649-y

[66] Yan Y et al. Probiotic *Bifidobacterium lactis* V9 attenuates hepatic steatosis and inflammation in rats with non-alcoholic fatty liver disease of article. AMB Express. 2020;**10**(1):101. DOI: 10.1186/s13568-020-01038-y

[67] Wang L et al. *Bifidobacterium bifidum* shows more diversified ways of relieving non-alcoholic fatty liver compared with Bifidobacterium adolescentis of article. Biomedicine. 2021;**10**:1. DOI: 10.3390/biomedicines10010084

[68] Zhang X et al. *Bifidobacterium longum* BL-19 inhibits oxidative stress and inflammatory damage in the liver of mice with NAFLD by regulating the production of butyrate in the intestine of article. Food Science & Nutrition. 2024;**12**(9):6442-6460. DOI: 10.1002/fsn3.4279

[69] Long X et al. *Bifidobacterium adolescentis* alleviates liver steatosis and Steatohepatitis by increasing fibroblast growth factor 21 sensitivity of article. Frontiers in Endocrinology (Lausanne). 2021;**12**:773340. DOI: 10.3389/fendo.2021.773340

[70] Lv H et al. In vitro probiotic properties of *Bifidobacterium animalis* subsp. lactis SF and its alleviating effect on non-alcoholic fatty liver disease of article. Nutrients. 2023;**15**:6. DOI: 10.3390/nu15061355

[71] Song Q et al. *Bifidobacterium pseudolongum*-generated acetate suppresses non-alcoholic fatty liver disease-associated hepatocellular carcinoma of article. Journal of Hepatology. 2023;**79**(6):1352-1365. DOI: 10.1016/j.jhep.2023.07.005

[72] Do MH et al. *Bifidobacterium animalis* ssp. lactis MG741 reduces body weight and ameliorates nonalcoholic fatty liver disease via improving the gut permeability and amelioration of inflammatory cytokines of article. Nutrients. 2022;**14**:9. DOI: 10.3390/nu14091965

[73] Zhao Z et al. *Lactobacillus plantarum* NA136 ameliorates nonalcoholic fatty liver disease by modulating gut microbiota, improving intestinal barrier integrity, and attenuating inflammation of article. Applied Microbiology and Biotechnology. 2020;**104**(12):5273-5282. DOI: 10.1007/s00253-020-10633-9

[74] Zhao C et al. *Lactobacillus acidophilus* YL01 and its exopolysaccharides ameliorate obesity and insulin resistance in obese mice via modulating intestinal specific bacterial groups and AMPK/ACC signaling pathway of article. International Journal of Biological Macromolecules. 2025;**300**:140287. DOI: 10.1016/j.ijbiomac.2025.140287

[75] Zhu L et al. *Lactobacillus salivarius* SNK-6 regulates liver lipid metabolism partly via the miR-130a-5p/MBOAT2 pathway in a NAFLD model of laying hens of article. Cells. 2022;**11**:24. DOI: 10.3390/cells11244133

[76] Wang Q et al. *Lactobacillus helveticus* R0052 alleviates liver injury by modulating gut microbiome and metabolome in D-galactosamine-treated rats of article. Applied Microbiology and Biotechnology. 2019;**103**(23-24):9673-9686. DOI: 10.1007/s00253-019-10211-8

[77] Huang L et al. *Lactobacillus plantarum* C88 protects against aflatoxin B(1)-induced liver injury in mice via inhibition of NF-kappaB-mediated inflammatory responses and excessive apoptosis of article. BMC Microbiology. 2019;**19**(1):170. DOI: 10.1186/s12866-019-1525-4

[78] Zhao Y, et al. Probiotic *Limosilactobacillus reuteri* DSM 17938 alleviates acute liver injury by activating the AMPK signaling via gut microbiota-derived propionate. Probiotics and Antimicrobial Proteins. 28 Jan 2025. DOI: 10.1007/s12602-025-10464-y [Epub ahead of print]

[79] Yang M et al. Preventive effect and molecular mechanism of lactobacillus rhamnosus JL1 on food-borne obesity in mice of article. Nutrients. 2021;**13**:11. DOI: 10.3390/nu13113989

[80] Wang Q et al. *Lactobacillus reuteri* CCFM8631 alleviates hypercholesterolaemia caused by the paigen atherogenic diet by regulating the gut microbiota of article. Nutrients. 2022;**14**:6. DOI: 10.3390/nu14061272

[81] Jo YM et al. *Lactobacillus gasseri* BNR17 and *Limosilactobacillus fermentum* ABF21069 ameliorate high sucrose-induced obesity and fatty liver via exopolysaccharide production and

beta-oxidation of article. Journal of Microbiology. 2024;**62**(10):907-918. DOI: 10.1007/s12275-024-00173-6

[82] Ting WJ et al. Heat killed *Lactobacillus reuteri* GMNL-263 reduces fibrosis effects on the liver and heart in high fat diet-hamsters via TGF-beta suppression of article. International Journal of Molecular Sciences. 2015;**16**(10):25881-25896. DOI: 10.3390/ijms161025881

[83] Lau HC et al. *Lactobacillus acidophilus* suppresses non-alcoholic fatty liver disease-associated hepatocellular carcinoma through producing valeric acid of article. eBioMedicine. 2024;**100**:104952. DOI: 10.1016/j.ebiom.2023.104952

[84] Mohseni AH et al. Modulation of the PI3K/Akt/mTOR signaling pathway by probiotics as a fruitful target for orchestrating the immune response of article. Gut Microbes. 2021;**13**(1):1-17. DOI: 10.1080/19490976.2021.1886844

[85] Deng Y et al. *Limosilactobacillus fermentum* TY-S11 ameliorates hypercholesterolemia via promoting cholesterol excretion and regulating gut microbiota in high-cholesterol diet-fed apolipoprotein E-deficient mice of article. Heliyon. 2024;**10**(11):e32059. DOI: 10.1016/j.heliyon.2024.e32059

[86] Shin JH et al. *Faecalibacterium prausnitzii* prevents hepatic damage in a mouse model of NASH induced by a high-fructose high-fat diet of article. Frontiers in Microbiology. 2023;**14**:1123547. DOI: 10.3389/fmicb.2023.1123547

[87] Munukka E et al. *Faecalibacterium prausnitzii* treatment improves hepatic health and reduces adipose tissue inflammation in high-fat fed mice of article. The ISME Journal. 2017;**11**(7):1667-1679. DOI: 10.1038/ismej.2017.24

[88] Hu W et al. Specific strains of *Faecalibacterium prausnitzii* ameliorate nonalcoholic fatty liver disease in mice in association with gut microbiota regulation of article. Nutrients. 2022;**14**:14. DOI: 10.3390/nu14142945

[89] Hong Y et al. *Desulfovibrio vulgaris*, a potent acetic acid-producing bacterium, attenuates nonalcoholic fatty liver disease in mice of article. Gut Microbes. 2021;**13**(1):1-20. DOI: 10.1080/19490976.2021.1930874

[90] Seo B et al. *Roseburia* spp. abundance associates with alcohol consumption in humans and its administration ameliorates alcoholic fatty liver in mice of article. Cell Host & Microbe. 2020;**27**(1):25-40.e6. DOI: 10.1016/j.chom.2019.11.001

[91] Fan M et al. Heat-killed enterococcus faecalis EF-2001 attenuate lipid accumulation in diet-induced obese (DIO) mice by activating AMPK signaling in liver of article. Food. 2022;**11**:4. DOI: 10.3390/foods11040575

[92] Lee JH et al. Heat-killed *Enterococcus faecalis* inhibit FL83B hepatic lipid accumulation and high fat diet-induced fatty liver damage in rats by activating lipolysis through the regulation the AMPK signaling pathway of article. International Journal of Molecular Sciences. 2023;**24**:5. DOI: 10.3390/ijms24054486

[93] Xu W et al. *Enterococcus faecium* GEFA01 alleviates hypercholesterolemia by promoting reverse cholesterol transportation via modulating the gut microbiota-SCFA axis of article. Frontiers in Nutrition. 2022;**9**:1020734. DOI: 10.3389/fnut.2022.1020734

[94] Streidl T et al. The gut bacterium *Extibacter muris* produces secondary

bile acids and influences liver physiology in gnotobiotic mice of article. Gut Microbes. 2021;**13**(1):1-21. DOI: 10.1080/19490976.2020.1854008

[95] Vajro P et al. Effects of *Lactobacillus rhamnosus* strain GG in pediatric obesity-related liver disease of article. Journal of Pediatric Gastroenterology and Nutrition. 2011;**52**(6):740-743. DOI: 10.1097/MPG.0b013e31821f9b85

[96] Abhari K et al. The effects of *Bacillus coagulans* supplementation in patients with non-alcoholic fatty liver disease: A randomized, placebo-controlled, clinical trial of article. Clinical Nutrition ESPEN. 2020;**39**:53-60. DOI: 10.1016/j.clnesp.2020.06.020

[97] Hsieh RH et al. *Bacillus coagulans* TCI711 supplementation improved nonalcoholic fatty liver by modulating gut microbiota: A randomized, placebo-controlled, clinical trial of article. Current Developments in Nutrition. 2024;**8**(3):102083. DOI: 10.1016/j.cdnut.2024.102083

[98] Ziegler MC et al. Impact of probiotic supplementation in a patient with type 2 diabetes on glycemic and lipid profile of article. Clinical Nutrition ESPEN. 2022;**49**:264-269. DOI: 10.1016/j.clnesp.2022.04.002

[99] Pedret A et al. Effects of daily consumption of the probiotic Bifidobacterium animalis subsp. lactis CECT 8145 on anthropometric adiposity biomarkers in abdominally obese subjects: A randomized controlled trial of article. International Journal of Obesity. 2019;**43**(9):1863-1868. DOI: 10.1038/s41366-018-0220-0

[100] Bai Z et al. Gut microbiome and metabolome alterations in overweight or obese adult population after weight-loss *Bifidobacterium* breve BBr60 intervention: A randomized controlled trial of article. International Journal of Molecular Sciences. 2024;**25**:20. DOI: 10.3390/ijms252010871

[101] Lee M et al. *Bifidobacterium lactis* IDCC 4301 (*B. Lactis* fit) supplementation effects on body fat, serum triglyceride, and adipokine ratio in obese women: A randomized clinical trial of article. Food & Function. 2024;**15**(16):8448-8458. DOI: 10.1039/d4fo00535j

[102] Lim S et al. Effect of *Lactobacillus sakei*, a probiotic derived from kimchi, on body fat in Koreans with obesity: A randomized controlled study of article. Endocrinology and Metabolism. (Seoul). 2020;**35**(2):425-434. DOI: 10.3803/EnM.2020.35.2.425

[103] Oh SJ et al. Effect of *Lactobacillus sakei* OK67 in reducing body and visceral fat in lifestyle-modified overweight individuals: A 12-week, randomized, double-blind. Placebo-Controlled Clinical Trial of article. Nutrients. 2023;**15**:13. DOI: 10.3390/nu15133074

[104] Sanchez M et al. Effect of *Lactobacillus rhamnosus* CGMCC1.3724 supplementation on weight loss and maintenance in obese men and women of article. The British Journal of Nutrition. 2014;**111**(8):1507-1519. DOI: 10.1017/S0007114513003875

[105] Tenorio-Jimenez C et al. *Lactobacillus reuteri* V3401 reduces inflammatory biomarkers and modifies the gastrointestinal microbiome in adults with metabolic syndrome: The PROSIR study of article. Nutrients. 2019;**11**:8. DOI: 10.3390/nu11081761

[106] Sohn M et al. Effect of *Lactobacillus plantarum* LMT1-48 on body fat in overweight subjects: A randomized, double-blind, placebo-controlled trial of

article. Diabetes and Metabolism Journal. 2023;**47**(1):92-103. DOI: 10.4093/dmj.2021.0370

[107] Kadooka Y et al. Regulation of abdominal adiposity by probiotics (*Lactobacillus gasseri* SBT2055) in adults with obese tendencies in a randomized controlled trial of article. European Journal of Clinical Nutrition. 2010;**64**(6):636-643. DOI: 10.1038/ejcn.2010.19

[108] Kim J et al. *Lactobacillus gasseri* BNR17 supplementation reduces the visceral fat accumulation and waist circumference in obese adults: A randomized, double-blind, placebo-controlled trial of article. Journal of Medicinal Food. 2018;**21**(5):454-461. DOI: 10.1089/jmf.2017.3937

[109] Nagata S et al. The effects of the *Lactobacillus casei* strain on obesity in children: A pilot study of article. Beneficial Microbes. 2017;**8**(4):535-543. DOI: 10.3920/BM2016.0170

[110] Choi BS et al. *Lacticaseibacillus rhamnosus* HA-114 improves eating behaviors and mood-related factors in adults with overweight during weight loss: A randomized controlled trial of article. Nutritional Neuroscience. 2023;**26**(7):667-679. DOI: 10.1080/1028415X.2022.2081288

[111] Alisi A et al. Randomised clinical trial: The beneficial effects of VSL#3 in obese children with non-alcoholic steatohepatitis of article. Alimentary Pharmacology & Therapeutics. 2014;**39**(11):1276-1285. DOI: 10.1111/apt.12758

[112] Wong VW-S et al. Treatment of nonalcoholic steatohepatitis with probiotics. A proof-of-concept study of article. Annals of Hepatology. 2013;**12**(2):256-262. DOI: 10.1016/s1665-2681(19)31364-x

[113] Kobyliak N et al. A multistrain probiotic reduces the fatty liver index, cytokines and aminotransferase levels in NAFLD patients: Evidence from a randomized clinical trial of article. Journal of Gastrointestinal and Liver Diseases. 2018;**27**(1):41-49. DOI: 10.15403/jgld.2014.1121.271.kby

[114] Mohamad Nor MH et al. The effect of probiotics (MCP((R)) BCMC((R)) strains) on hepatic steatosis, small intestinal mucosal immune function, and intestinal barrier in patients with non-alcoholic fatty liver disease of article. Nutrients. 2021;**13**:9. DOI: 10.3390/nu13093192

[115] Crommen S et al. A specifically tailored multistrain probiotic and micronutrient mixture affects nonalcoholic fatty liver disease-related markers in patients with obesity after mini gastric bypass surgery of article. The Journal of Nutrition. 2022;**152**(2):408-418. DOI: 10.1093/jn/nxab392

[116] Escouto GS et al. Probiotic supplementation, hepatic fibrosis, and the microbiota profile in patients with nonalcoholic Steatohepatitis: A randomized controlled trial of article. The Journal of Nutrition. 2023;**153**(7):1984-1993. DOI: 10.1016/j.tjnut.2023.05.019

[117] Sepideh A et al. Effects of multistrain probiotic supplementation on glycemic and inflammatory indices in patients with nonalcoholic fatty liver disease: A double-blind randomized clinical trial of article. Journal of the American College of Nutrition. 2016;**35**(6):500-505. DOI: 10.1080/07315724.2015.1031355

[118] Ahn SB et al. Randomized, double-blind, placebo-controlled study of a multispecies probiotic mixture in nonalcoholic fatty liver disease of article.

Scientific Reports. 2019;**9**(1):5688. DOI: 10.1038/s41598-019-42059-3

[119] Solito A et al. Supplementation with bifidobacterium breve BR03 and B632 strains improved insulin sensitivity in children and adolescents with obesity in a cross-over, randomized double-blind placebo-controlled trial of article. Clinical Nutrition. 2021;**40**(7):4585-4594. DOI: 10.1016/j.clnu.2021.06.002

[120] Kwon HS et al. The effect of the *Lacticaseibacillus paracasei* BEPC22 and *Lactiplantibacillus plantarum* BELP53 combination (BN-202M) on body fat percentage loss in overweight individuals: A randomized, double-blind. Placebo-Controlled Study of Article. Nutrients. 2024;**16**:13. DOI: 10.3390/nu16131993

[121] Mo SJ et al. Effects of lactobacillus curvatus HY7601 and lactobacillus plantarum KY1032 on overweight and the gut microbiota in humans: Randomized, double-blinded. Placebo-Controlled Clinical Trial of Article. Nutrients. 2022;**14**:12. DOI: 10.3390/nu14122484

[122] Zikou E et al. The effect of probiotic supplements on metabolic parameters of people with type 2 diabetes in Greece-A randomized, double-blind. Placebo-Controlled Study of Article. Nutrients. 2023;**15**:21. DOI: 10.3390/nu15214663

[123] Kobyliak N et al. Effect of alive probiotic on insulin resistance in type 2 diabetes patients: Randomized clinical trial of article. Diabetes and Metabolic Syndrome: Clinical Research and Reviews. 2018;**12**(5):617-624. DOI: 10.1016/j.dsx.2018.04.015

[124] Mofidi F et al. Synbiotic supplementation in lean patients with non-alcoholic fatty liver disease: A pilot, randomised, double-blind, placebo-controlled, clinical trial of article. The British Journal of Nutrition. 2017;**117**(5):662-668. DOI: 10.1017/S0007114517000204

[125] Manzhalii E et al. Treatment efficacy of a probiotic preparation for non-alcoholic steatohepatitis: A pilot trial of article. Journal of Digestive Diseases. 2017;**18**(12):698-703. DOI: 10.1111/1751-2980.12561

[126] Sayari S et al. Combined effects of synbiotic and sitagliptin versus sitagliptin alone in patients with nonalcoholic fatty liver disease of article. Clinical and Molecular Hepatology. 2018;**24**(3):331-338. DOI: 10.3350/cmh.2018.0006

[127] Eslamparast T et al. Synbiotic supplementation in nonalcoholic fatty liver disease: A randomized, double-blind, placebo-controlled pilot study of article. The American Journal of Clinical Nutrition. 2014;**99**(3):535-542. DOI: 10.3945/ajcn.113.068890

[128] Baba Y et al. Effect of intake of Bifidobacteria and dietary fiber on resting energy expenditure: A randomized, placebo-controlled, double-blind, parallel-group comparison study of article. Nutrients. 2024;**16**:14. DOI: 10.3390/nu16142345

[129] Lauw S et al. Effects of synbiotic supplementation on metabolic syndrome traits and gut microbial profile among overweight and obese Hong Kong Chinese individuals: A randomized trial of article. Nutrients. 2023;**15**:19. DOI: 10.3390/nu15194248

[130] Rondanelli M et al. Effects of 60-day *Saccharomyces boulardii* and superoxide dismutase supplementation on body composition, hunger sensation, pro/antioxidant ratio, inflammation and hormonal Lipo-metabolic biomarkers in obese adults: A double-blind,

placebo-controlled trial of article. Nutrients. 2021;**13**:8. DOI: 10.3390/nu13082512

[131] Ferolla SM et al. Beneficial effect of synbiotic supplementation on hepatic steatosis and anthropometric parameters, but not on gut permeability in a population with nonalcoholic steatohepatitis of article. Nutrients. 2016;**8**:7. DOI: 10.3390/nu8070397

[132] Xiong K et al. *Lactobacillus casei* alleviated the abnormal increase of cholestasis-related liver indices during tuberculosis treatment: A post hoc analysis of randomized controlled trial of article. Molecular Nutrition & Food Research. 2021;**65**(16):e2100108. DOI: 10.1002/mnfr.202100108

[133] Chong PL et al. A randomised placebo controlled trial of VSL#3((R)) probiotic on biomarkers of cardiovascular risk and liver injury in non-alcoholic fatty liver disease of article. BMC Gastroenterology. 2021;**21**(1):144. DOI: 10.1186/s12876-021-01660-5

[134] Engel S et al. Safety of *Bifidobacterium* breve, Bif195, employing a human exercise-induced intestinal permeability model: A randomised, double-blinded, placebo-controlled, parallel group trial of article. Beneficial Microbes. 2022;**13**(3):243-252. DOI: 10.3920/BM2021.0173